Applied Mathematical Sciences
Volume 85

T0189768

Applied Mathematical Sciences

(continued following index)

Huijun Yang

Wave Packets and Their Bifurcations in Geophysical Fluid Dynamics

With 89 Figures

Springer-Verlag
New York Berlin Heidelberg London
Paris Tokyo Hong Kong Barcelona

Huijun Yang
Geophysical Fluid Dynamics Institute
The Florida State University
Tallahassee, FL 32306-3017
USA

Editors

F. John
Courant Institute of
  Mathematical Sciences
New York University
New York, NY 10012
USA

J.E. Marsden
Department of
  Mathematics
University of California
Berkeley, CA 94720
USA

L. Sirovich
Division of
  Applied Mathematics
Brown University
Providence, RI 02912
USA

Mathematics Subject Classifications: 86A35, 76C20

Library of Congress Cataloging-in-Publication Data
Yang, Huijun.
  Wave packets and their bifurcations in geophysical fluid dynamics/
  Huijun Yang.
    p. cm.—(Applied mathematical sciences; v. 85)

    1. Fluid dynamics. 2. Wave Packets. 3. Bifurcations. 4. Geophysics. I. Title.
  II. Series: Applied mathematical sciences (Springer-Verlag New York,
  Inc.); v. 85.
  QC809.F5Y26   1990
  550—dc20   90-31403

Photocomposed copy prepared using LaTeX.

Printed in the United States of America.

9 8 7 6 5 4 3 2 1

ISBN 978-1-4419-3093-4

Preface

The material in this book is based predominantly on my recent work. It is the first monograph on the subject, though some support material may overlap other monographs. The investigation of wave packets and their bifurcations is very interesting, and useful theoretically and in practice, not only in geophysical fluid dynamics, which is the field to which the theory is being applied here, but also in other fields in mathematics and the natural sciences. I hope that the applied mathematician will find reading this book worthwhile, especially the material on the behavior of highly nonlinear dynamic systems. However, it is my belief that applying the concepts and methods developed here to other fields will be both interesting and constructive, since there are numerous phenomena in other areas of physics that share the characteristics of those in geophysical fluid dynamics. The theory developed here provides an effective tool to investigate the structure and the structural changes of dynamic systems in physics. Applications of the theory in geophysical fluid dynamics are an example of its usefulness and effectiveness. Some of the results presented here give us more insight into the nature of geophysical fluids.

Moreover, the material is presented systematically and developmentally. Necessary basic knowledge is provided to make the book more readable for graduate students and researchers in such fields as applied mathematics, geophysical fluid dynamics, atmospheric sciences, and physical oceanography. The book, therefore, can also be used as a reference for or a supplement in courses like wave theory, bifurcation theory, advanced dynamic meteorology, advanced dynamic oceanography, and geophysical fluid dynamics.

The book has eight chapters. Chapter 1 is the shortest one and presents some background in geophysical fluid dynamics, from the basic laws to various approximations of such equations as the Boussinesq approximation, the earth's f-plane and β-plane approximations, and the earth's δ-surface approximation. In Chapter 2 the theory of wave packet is developed systematically. The chapter begins with the wave packet representation and then discusses the asymptotic solution of wave equations and the WKB method and their relationship. General theorems of the wave packet are also addressed, including an example of a Rossby wave packet, the eikonal equation, the theory in stationary and moving media, wave-action conservation, and the energy exchange with basic flow. The generalized Lagrangian mean description developed by Andrews and McIntyre and general wave-action in this description are presented in Section 2.6.

Chapters 3 to 6 are devoted to discussions of the evolution of the wave packet in barotropic basic flows, with the emphasis on the structure and structural change of the wave packet. Chapter 3 introduces the inviscid shallow-water model and potential vorticity equation. Equations governing the evolution of the Rossby wave packet are derived. Integral properties of a wave packet are discussed, including extensions of the Rayleigh theorem and the Fjørtoft theorem. The effects of the role of the earth's rotation, basic flow, and topography upon the structure and structural change of a wave packet are covered. Chapter 4 introduces the concept of the wave packet's structural vacillation (or oscillation) and considers its relevance to vacillations observed in geophysical fluids. The δ-effect is also discussed. Chapters 5 and 6 consider the bifurcation behavior of wave packets by the use of bifurcation diagrams and phase space diagrams. The dynamic behavior of the structure of a wave packet is discussed. The dynamic system consists of two highly nonlinear equations, (5.26) and (5.27). Topics such as equilibrium state, stability, topological structure near an equilibrium state, primary bifurcation, secondary bifurcation, cascading bifurcation, and structural stability are also discussed, as required.

Chapter 7 extends earlier studies in barotropic basic flows to the case of three-dimensional wave packets in stratified baroclinic basic flows. Most of the results in this chapter have not been published before. The structures of developing and decaying wave packets and bifurcation properties are investigated in such cases as the stably stratified and unstably stratified purely barotropic basic flows, the purely baroclinic basic flows, and the barotropic and baroclinic basic flows. The dynamic system governing the structural change of a wave packet consists of three highly nonlinear equations (7.92) to (7.94). The bifurcation properties in both stably stratified and unstably stratified basic flows are investigated. The implications of their results in geophysical fluids are discussed.

The last chapter deals with the theory of wave packet propagation associated with teleconnections observed in geophysical fluids. The theory on a sphere, developed by Hoskins and Karoly, is also discussed. In contrast to Chapters 3 to 7, it emphasizes the propagation property of a wave packet rather than the structures and structural changes. Hence, the discussions in Chapters 3 to 7 and those in Chapter 8 can support each other under the main frame developed in Chapters 1 and 2. The entire content of the manuscript, therefore, is consistent and carefully organized.

I would like to thank the many people who helped in my study and the present work. I especially thank those people who are in the Geophysical Fluid Dynamics Institute of Florida State University, Peking University, the Institute of Atmospheric Physics of Academia Sinica, and the Chegdu Institute of Meteorology. I am very grateful to Professor Dasheng Yang and Professor Ronghui Huang for their continuing encouragement and for their review of the manuscript. Dr. Arthur J. Rosenthal read the manuscript and offered very helpful suggestions. Dr. Xiaolei Zou read Chapter 7 and helped me in the final draft. To them I am sincerely grateful. Dr. Phillips G. Drazin's suggestions were also very helpful. I am indebted to Dr. Richard L. Pfeffer, Director of the Geophysical Fluid Dynamics Institute, who has given me continuing encouragement and support during the course of the present work. My sincere thanks are also due to Drs. Lious N. Howard, Ruby Krishnamurti, and Henyi Weng for their encouragement and help. During the writing, I also received help from Drs. Brian J. Hoskins, Rolando R. Garcia, and David J. Karoly.

I am also indebted to Mr. Bernd Grossmann, Vice-President of Springer-Verlag Heidelberg, Inc., Mr. Peng Songjian, Chief Editor of Peking University Press, Ms. Qiu Shuqing, Director of Mathematics and Physics of Peking University Press, as well as Ms. Wendy Rice and Ms. Francine Sikorski at Springer-Verlag New York, Inc., for their splendid cooperation.

Finally, I am particularly appreciative of my wife, Dr. Liwen Tao, for her considerable patience and help. I also owe thanks to my little daughter, Linda Fei Yang.

This work was supported in part by the Geophysical Fluid Dynamics Institute of Florida State University. I gratefully acknowledge this support. I also thank the Florida State University Chapter of The Society of Sigma Xi and Peking University for their awards, given for my research papers on the subject. This is Contribution Number 285 of the Geophysical Fluid Dynamics Institute.

Tallahassee, Florida   *Huijun Yang*
Spring 1990

Contents

1

Introduction

1.1 The Nature of Geophysical Fluids and Geophysical Fluid Dynamics

Geophysical fluids, as considered here, mean the fluids in which the dynamics of the earth's rotation, that is, the Coriolis force and gravitational force, play very important roles. These properties distinguish them from general fluids and make them unique in the field of fluid dynamics.

Geophysical fluid dynamics is about fundamental dynamic concepts and processes essential to an understanding of the nature of geophysical fluids. The geophysical fluids we are studying here are the atmosphere and the oceans. Moreover, we are focusing on the large-scale motions the atmospheric and oceanic dynamics evidently have in common. The study of one often enriches our understanding of the other.

Other important characteristics of geophysical fluid dynamics include nonlinearity due to advection, differential heating due to the difference between the solar radiation received and the heat capacity of the earth's lands and oceans, dissipation due to the viscosity of the fluid during molecular and turbulent motion, and air–sea interaction. For instance, it is found that oceanic currents on the surface are mainly driven by the stress of motions above the oceans; thus they are called *wind-driven ocean circulations.*

1.2 The Basic Equations

Motion in geophysical fluids is governed by the conservation laws for mass, momentum, and equations of state and by the laws of thermodynamics. Using an Enlerian description of the motion, we treat the particle velocity **u**, pressure p, density ρ, temperature T, and salinity S (only in oceans) as functions of the position vector **r**, which is measured outward from the earth's center, and the time t. All positions are referred to a right-handed, orthogonal coordinate system, which is uniformly rotating with the earth's angular velocity $\boldsymbol{\Omega}$, with a magnitude of

$$\Omega = |\boldsymbol{\Omega}| = 7.29 \times 10^{-5}\,\mathrm{rad/s}.$$

The velocity $\mathbf{u}_r$ in this rotating frame of reference is related to the inertial velocity, that is, the absolute velocity $\mathbf{u}_a$, by the equation

$$\mathbf{u}_a = \mathbf{u}_r + \mathbf{\Omega} \times \mathbf{r}. \tag{1.1}$$

The conservation of mass: *Continuity equation.* In the absence of sources or sinks of mass within the fluid, the condition of mass conservation is expressed by the continuity equation

$$\frac{\partial \rho}{\partial t} + \nabla \cdot (\rho \mathbf{u}) = 0, \tag{1.2}$$

where $\mathbf{u}$ is the velocity in the rotating frame.

This equation states that the local increase of density with time must be balanced by a divergence of mass flux, $\rho \mathbf{u}$. The equation can also be written

$$\frac{d\rho}{dt} + \rho \nabla \cdot \mathbf{u} = 0, \tag{1.3}$$

where

$$\frac{d}{dt} \equiv \frac{\partial}{\partial t} + \mathbf{u} \cdot \nabla \tag{1.4}$$

is the total derivative (or the substantial derivative) with respect to the time of any property following individual fluid elements.

The momentum conservation. The momentum conservation is expressed as

$$\rho \frac{d\mathbf{u}}{dt} + \rho 2\mathbf{\Omega} \times \mathbf{u} = \rho[\mathbf{g} - \mathbf{\Omega} \times (\mathbf{\Omega} \times \mathbf{r})] - \nabla p + \mathbf{F}, \tag{1.5}$$

where $\mathbf{g} = -g\hat{z}$ denotes the gravitational acceleration, $g = 9.8\,\text{m/s}^2$, $\hat{z}$ denotes a unity vector in the vertical direction, that is, outward from the center of the earth, and $\mathbf{F}$ represents the sum of all the other forces per unit volume acting on the fluid, including the tide-producing forces, as well as molecular and turbulent friction. The magnitude of the ratio of the centrifugal term $\rho \mathbf{\Omega} \times (\mathbf{\Omega} \times \mathbf{r})$ to the gravitational term $\rho \mathbf{g}$ is less than 3×10^{-3} in atmospheric and oceanic motion. Therefore, the term $\rho \mathbf{\Omega} \times (\mathbf{\Omega} \times \mathbf{r})$ can be neglected, so that the effects of rotation on the geophysical fluid are manifested only through the Coriolis term $\rho \mathbf{\Omega} \times \mathbf{u}$.

In addition to pressure and temperature, for oceans, we require the salinity S in order to specify a thermodynamic state, where S (in parts per thousand) is loosely defined as the mass in grams of all dissolved solids in 1 kg of seawater [for a more precise definition, see Fofonoff (1962)]. The density of seawater is then given by an equation of state, of the form

$$\rho = \rho(p, S, T). \tag{1.6}$$

This relation, in general, is nonlinear in p, T, and S and has no simple analytic form (Mamayev, 1975). However, in *double diffusive convection* problems (i.e., thermohaline or thermosolute convection), the following form is found in the literature (Turner, 1973):

$$\rho = \rho_0[1 - \alpha(T - T_0) + \beta(S - S_0)], \tag{1.7}$$

where ρ_0, T_0, and S_0 are mean values of density, temperature, and salinity, respectively; α and β are the coefficients of expansion due to the change in temperature and in salinity with respect to density, respectively.

For the atmosphere, the equation of state can be stated as

$$\rho = \frac{p}{RT}, \tag{1.8}$$

where R is the gas constant for the air, which is $287\ J\ \mathrm{deg}^{-1}\ \mathrm{kg}^{-1}$ for the dry air.

To close the above equations for seven unknown functions $\mathbf{u}$ (which has three components), p, ρ, T, and S, we need two more conservation equations for T and S.

The conservation of internal energy. The conservation of internal energy may be expressed by the following equation (Batchelor, 1967):

$$\frac{d(\rho C_v T)}{dt} = \nabla \cdot (K_T \nabla T) + Q_T, \tag{1.9}$$

where C_v denotes the specific heat at constant volume, K_T denotes the thermal conductivity, and Q_T represents all sources and sinks of heat. In particular, Q_T includes heating due to compression, cooling due to shearing motions, and all heat transfer terms (solar heating, evaporative cooling, sensible heat flux, long-wave radiation, etc.).

The conservation of salt. The conservation of salt in the ocean can be stated as follows:

$$\frac{dS}{dt} = \nabla \cdot (K_S \nabla S) + Q_S, \tag{1.10}$$

where K_S denotes the coefficient of molecular diffusion of salt and Q_S includes all sources and sinks of salt due to such phenomena as ice melting and ice formation, precipitation, and evaporation.

1.3 Approximations to the System

We have now obtained the closed set of equations that govern the motion of geophysical fluids. They are eqs. (1.2) or (1.3), (1.5), (1.6) or (1.8), (1.9), and (1.10). The governing equations, however, are just too complicated to solve them analytically. Thus, several approximations of the system are often used, either separately or in combination, to simplify the problem when different questions are asked. Here, we will discuss four kinds of approximations. They are usually known as the Boussinesq approximation, the f-plane approximation, the β-plane approximation, and the δ-surface approximation. The δ-surface approximation was recently introduced by Yang (1987, 1988). Each of these approximations imposes important physical and/or geometrical restrictions on the problem.

1.3.1   THE BOUSSINESQ APPROXIMATION

Boussinesq (1903) noticed that there are many situations of practical occurrence in which the basic equations can be simplified considerably. These situations occur when the variation in density is relatively small. Variability due to changes in temperature is only moderate. For changes in temperature not exceeding $10°$, say, the variations in density are at most 1%. Therefore, such small variations can, in general, be ignored. However, there is one important exception, the variability of ρ in the gravity force in the equation of motion cannot be ignored. Accordingly, we may treat the density ρ as a constant in all terms in the equations of motion, except the buoyancy force. This is the *Boussinesq approximation*. Although this approximation is named after Boussinesq, it was used earlier by Oberbeck (1888) in his study of the atmospheric Hadley Regime. Additional discussions on the Boussinesq approximation can be found in such work as that by Spiegel and Veronis (1960) and by Mihaljan (1962).

Therefore, the equations of motion in the form of the Boussinesq approximation are as follows.

The equation of continuity (1.3) will take the form

$$\nabla \cdot \mathbf{u} = 0. \tag{1.11}$$

The momentum equation (1.5), the form

$$\frac{d\mathbf{u}}{dt} + 2\mathbf{\Omega} \times \mathbf{u} = -\frac{1}{\rho_0}\nabla p + \frac{\rho}{\rho_0}\mathbf{g} + \frac{1}{\rho_0}\mathbf{F}. \tag{1.12}$$

The other equations are unchanged.

1.3.2   THE f-PLANE AND THE β-PLANE APPROXIMATION

We are primarily interested in motions in geophysical fluids of limited horizontal extent, especially for oceanic motion. Though the best way to describe motions in geophysical fluids is by a spherical polar coordinate system, it is often difficult to deal with in this system practice. Therefore, it is desirable to introduce a Cartesian metric centered at some reference latitude and longitude that locally approximates the spherical metric in

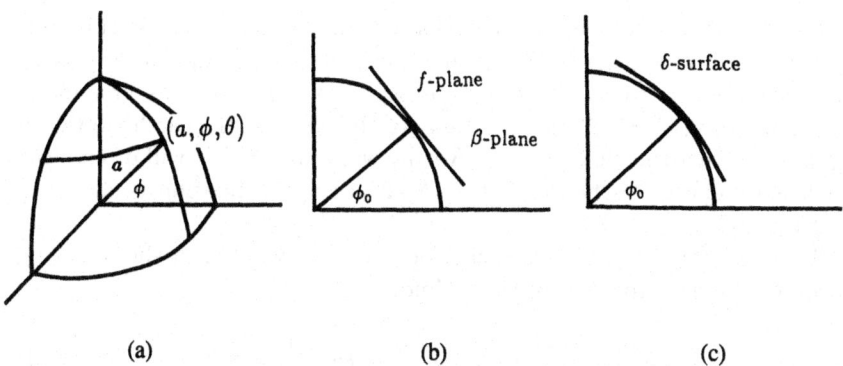

FIGURE 1.1. Approximations of systems on the earth's surface: (a) spherical coordinates, (b) the f-plane and the β-plane approximations, and (c) the δ-surface approximation. a is the average radius of the earth, ϕ is the latitude, and ϕ_0 is the reference latitude.

its chosen neighborhood of the earth's surface. We define a very important parameter, f, as the local vertical (or radial) component of 2Ω, that is,

$$f = 2\Omega \sin \phi, \qquad (1.13)$$

where ϕ is the latitude of the earth. This parameter is called the *Coriolis parameter*. In the metric, when the value of f is approximately constant, the *f-plane approximation* results. However, in this metric, the effects of sphericity are retained by approximating f with the linear function of y; this approximation is called the *β-plane approximation*. Here, y is the latitudinal coordinate and is measured positive northward from the reference latitude. This approximation, first introduced by Rossby and collaborators (1939) in his study of zonal circulation in the middle latitudes of the atmosphere, resulted in the well-known Rossby wave theory. With the β-plane approximation, the propagation properties of large-scale motions in geophysical fluids, especially those in the middle latitudes of the atmosphere, can be explained rationally. The name reflects the traditional use of the symbol β, which denotes $\partial f / \partial y$ evaluated at the reference point. Many studies have used the Rossby wave theory with the β-plane approximation. The f-plane and the β-plane approximations are illustrated in Figure 1.1; the spherical coordinate system is shown in (a), and the f-plane and the β-plane approximations are shown in (b).

1.3.3 THE δ-SURFACE APPROXIMATION

However, the β-plane approximation can only express the effect of the earth's rotation upon the propagation properties of the motion. To consider the effect of rotation upon the structural changes in the motions, in term of the Rossby wave, for instance, the second derivative of the Coriolis parameter with respect to latitude must be taken into account. In this metric, the effect of sphericity are retained by a quadratic function of y. This approximation is called the *δ-surface approximation* and was first introduced by Yang (1987, 1988) in a study of the effect of the earth's rotation upon the structural change in the Rossby wave packet. We will discuss the so-called δ-effect in Chapter 3. The δ-surface approximation is shown in Figure 1.1c.

The equations in these approximations can be written in the following form for the incompressive inviscid fluid.

$$\frac{\partial u}{\partial t} + u\frac{\partial u}{\partial x} + v\frac{\partial u}{\partial y} + w\frac{\partial u}{\partial z} - fv + \frac{1}{\rho}\frac{\partial p}{\partial x} = 0, \tag{1.14}$$

$$\frac{\partial v}{\partial t} + u\frac{\partial v}{\partial x} + v\frac{\partial v}{\partial y} + w\frac{\partial v}{\partial z} + fu + \frac{1}{\rho}\frac{\partial p}{y} = 0, \tag{1.15}$$

$$\frac{\partial w}{\partial t} + u\frac{\partial w}{\partial x} + v\frac{\partial w}{\partial y} + w\frac{\partial w}{\partial z} + \frac{1}{\rho}\frac{\partial p}{\partial z} + g = 0, \tag{1.16}$$

and

$$\frac{\partial u}{\partial x} + \frac{\partial v}{\partial y} + \frac{\partial w}{\partial z} = 0, \tag{1.17}$$

$$\frac{\partial \rho}{\partial t} + u\frac{\partial \rho}{\partial x} + v\frac{\partial \rho}{\partial y} + w\frac{\partial \rho}{\partial z} = 0, \tag{1.18}$$

where terms associated with the horizontal components of 2Ω and curvature terms have been ignored. Here, u, v, and w are velocity components in the x-, y-, and z-directions, respectively; x, y, and z are coordinates eastward, northward, and vertically, respectively.

For the f-plane approximation

$$f = f_0 = \text{constant}, \tag{1.19}$$

where

$$f_0 = 2\Omega \sin \phi_0. \tag{1.20}$$

For the β-plane approximation,

$$f = f_0 + \beta y, \tag{1.21}$$

where

$$\beta = \left(\frac{\partial f}{\partial y}\right)_{y=y_0} \tag{1.22}$$

$$= \frac{2\Omega \cos \phi_0}{a}, \tag{1.23}$$

and where a is the average radius of the earth.

For the δ-surface approximation,

$$f = f_0 + \beta y - \frac{1}{2}\delta y^2, \tag{1.24}$$

where

$$\delta = \frac{2\Omega \sin \phi_0}{a^2}. \tag{1.25}$$

1.4 Closure

Readers who want to know more about the fundamentals of geophysical fluid dynamics are encouraged to consult the book by Pedlosky (1987). There are excellent discussions about the basic theory of the subject. Readers may also find more fundamental discussions in a book by Holton (1979), which emphasizes application in the atmosphere, and a book by Pond and Pickard (1983), which focuses mainly on physical oceanography. The book by Gill (1982) is also highly recommended.

REFERENCES

Batchelor, G.K. (1967). *An Introduction to Fluid Dynamics.* Cambridge University Press, London.

Boussinesq, J. (1903). Théorie analytique de la chaleaur (Paris: Gauthier-Vellars). **2**, 172.

Fofonoff, N.P. (1962). The physical properties of sea waters. *The Sea Interscience* **1**, 3–30.

Gill, A.E. (1982). *Atmosphere–Ocean Dynamics.* Academic Press, New York, London.

Holton, J.R. (1979). *An Introduction to Dynamical Meteorology.* 2nd ed. Academic Press, New York.

Mamayev, O.I. (1975). *Temperature–Salinity Analysis of World Oceans Waters.* Elsvier, Amsterdam.

Mihaljan, J.M. (1962). A virogous exposition of the Boussinesq approximations applicable to a thin layer of fluid. *Astrophysical J.* **136**, 1126–1133.

Oberbeck, A. (1888). Über die Bewegungserscheinungen der Atmosphere. Sitzb. K. Presuss. Akad. Wiss. pp. 383–395 and pp. 1129–1138. Translated by C. Abbe in *Smithsonian Miss. Coll.* (1891).

Pedlosky, J. (1987). *Geophysical Fluid Dynamics.* 2nd ed. Springer-Verlag, New York.

Pond, S., and Pickard, G.L. (1983). *Introduction to Dynamical Oceanography.* 2nd ed. Pergamon Press, Toronto, New York.

Rossby, C.-G., and collaborators (1939). Relation between variations in the intensity of the zonal circulation of the atmosphere and the displacements of the semi-permanent centers of action. *J. Mar. Res.* **2**, 38–55.

Spiegel, E.A., and Veronis, G. (1960). On the Boussinesq approximation for a compressible fluid. *Astrophysical J.* **131**, 442–447.

Turner, J.S. (1973). *Buoyancy Effects in Fluids.* Cambridge University Press, London.

Yang, H. (1987). Evolution of a Rossby wave packet in barotropic flows with asymmetric basic current, topography and δ-effect. *J. Atmos. Sci.* **44**, 2267–2276.

Yang, H. (1988). Global behavior of the evolution of a Rossby wave packet in barotropic flows on the earth's δ-surface. *J. Atmos. Sci.* **45**, 133–146.

2

The Wave Packet Theory

2.1 Introduction

In this chapter we discuss the basic theory of the wave packet, beginning with a representation of an arbitrary disturbance system in the form of the wave packet. It will be shown that the single wave can be considered a special case of the wave packet. Hence, the wave theory, in general, can be considered the basic ingredient of the wave packet theory. In Section 2.3, the asymptotic behavior of wave is discussed and it is to be consistent with the wave packet representation of the disturbance system. That is, the energy of the disturbance system propagates at the group velocity. The WKB approximation is another way of looking at the disturbance system in the form of the wave packet. This material is addressed in Section 2.4.

Section 2.5 covers the main part of the theory, which is called the *general theory*. This theory is also called the *ray theory* or the *kinematic theory* in the literature. A very important theorem, *the structure independence theorem* of the wave packet is discussed; it has not been considered before. The theorem can be applied to most practical problems. Moreover, the theorem simplifies the problem considerably and allows us more insight into the problem. Other theorems about the wave packet are also addressed. The reader should become familiar with the theory, since the rest of the chapters are based upon it.

In Section 2.6, we introduce a new development in the wave theory with the generalized Lagrangian-mean (GLM) description. This description, developed by Andrews and McIntyre (1978b,c); has had a great impact on studies of geophysical fluids, especially those of the atmosphere. However, we only cover the basics of the theory, specifically: the GLM description and the general wave-action equation with the basic theorems of mean-flow (Appendix A).

2.2 The Wave Packet Representation for an Arbitrary Disturbance System

2.2.1 THE CONCEPT OF WAVE PACKET AND GROUP VELOCITY

Suppose there are two single wave

$$\phi_1(x,t) = A_0 \cos m_1(x - c_1 t) \tag{2.1}$$

and

$$\phi_2(x,t) = A_0 \cos m_2(x - c_2 t), \tag{2.2}$$

where m_1 and m_2 are wave numbers, c_1 and c_2 are the phase speeds of the waves, and A_0 is the amplitude of the waves. Then, the two waves can consist of a group wave:

$$\begin{aligned} \phi(x,t) &= \phi_1(x,t) + \phi_2(x,t) \\ &= A(x,t) \cos m_0(x - c_0 t), \end{aligned} \tag{2.3}$$

where

$$A(x,t) = 2A_0 \cos \frac{\Delta m}{2} \left[x - \left(\frac{\partial mc}{\partial m} \right)_0 t \right] \tag{2.4}$$

and

$$\Delta m = m_2 - m_1, \tag{2.5a}$$

$$m_0 = \frac{m_1 + m_2}{2}, \tag{2.5b}$$

and

$$c_0 = c(m_0) = \frac{c_1 + c_2}{2}, \tag{2.5c}$$

provided

$$\left| \frac{\Delta m}{m_0} \right| \ll 1.$$

This group wave consists of two parts. The first part is $\cos m_0(x - c_0 t)$ and the second part is the amplitude wave $A(x,t)$, which is a maximum line of the group wave. Usually $A(x,t)$ varies slowly with respect to space and time. This group wave is called a *wave packet*; it is shown in Figure 2.1. The wave packet has a phase velocity, c_0, and a wave number, m_0. For the small amplitude wave, the wave energy is proportional to the square of the amplitude, and from eq. (2.4), one finds that the energy of the wave packet propagates along the velocity $(\partial mc/\partial m)$. This velocity is called the *group velocity*. In the following sections, we develop a method to represent an arbitrary perturbation in the form of the wave packet.

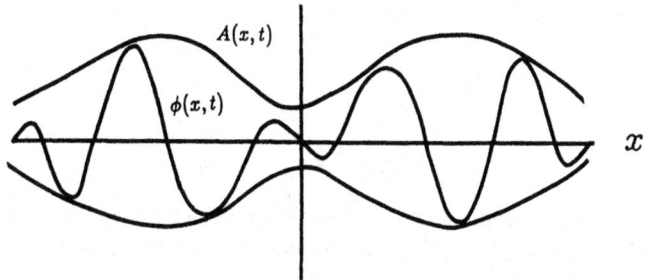

FIGURE 2.1. Concept of the wave packet.

(1) Physical Intuition

As we know, a single wave is usually not sufficient to describe a disturbance system. In geophysical flows, we often observe that weather systems (or oceanic wave patterns) cannot be described by a single wave, nor can other fields associated with propagation problems, in general. However, it is true that there are always some dominant waves in the disturbance system, which can easily be seen from the wave number spectrum. For instance, in the winter, the high level atmosphere is often observed to be dominated by waves with the numbers two and four. Therefore, it is reasonable to approximate the disturbance system with the dominant wave or waves.

From a mathematical point of view, it can be proven that this approximation theory is valid generally, and that the wave packet propagates at the group velocity of the dominant wave. From the Fourier theory, an arbitrary disturbance system can be always represented, as follows:

$$\phi(\mathbf{r}, t) = \int \int \int \Phi_{m,n,k}(t)e^{i\mathbf{l}\cdot\mathbf{r}}dmdndk, \qquad (2.6)$$

where $\mathbf{l}$ is the wave number vector with components m, n, and k. The integration is taken over the whole space of the wave number spectrum. Here, $\Phi_{m,n,k}e^{i\mathbf{l}\cdot\mathbf{r}}$ is a single wave, and can be written as

$$\Phi_{m,n,k}(t)e^{i\mathbf{l}\cdot\mathbf{r}} = \Psi_{m,n,k}e^{i(\mathbf{l}\cdot\mathbf{r}-\sigma t)}, \qquad (2.7)$$

where σ is its frequency. Hence, an arbitrary disturbance system can be written as

$$\phi = \int \int \int \Psi_{m,n,k}e^{i(\mathbf{l}\cdot\mathbf{r}-\sigma t)}dmdndk. \qquad (2.8)$$

Again, the integration is taken over the whole space of the wave number spectrum. The frequency is, in general, a function of the wave number and physical medium parameters.

The *phase velocity* of the wave is defined as

$$\mathbf{C} = \mathbf{i}\frac{\sigma}{m} + \mathbf{j}\frac{\sigma}{n} + \mathbf{k}\frac{\sigma}{k}, \qquad (2.9)$$

and the *group velocity* is defined by

$$\mathbf{C}_g = \frac{\partial \sigma}{\partial \mathbf{l}}$$

$$= \mathbf{i}\frac{\partial \sigma}{\partial m} + \mathbf{j}\frac{\partial \sigma}{\partial n} + \mathbf{k}\frac{\partial \sigma}{\partial k}. \tag{2.10}$$

A wave is said to be *nondispersive* if its phase velocity is independent of the wave number. A wave is said to be *dispersive* if its phase velocity depends on the wave number.

Consider two cases. In the first case, suppose there is a single dominant peak in the wave number spectrum. In the second case, there are N dominant peaks in the wave number spectrum.

2.2.2 CASE I: A SINGLE PEAK IN THE WAVE NUMBER SPECTRUM

This case is illustrated in Figure 2.2. In the figure, l is the total wave number of the disturbance system, and $\mathbf{\Psi}_{m,n,k}$ is the maximum spectrum at each l. Figure 2.2a illustrates the discrete spectra, and Figure 2.2b is a typical distribution of a continuous spectrum. From these figures, we find that the dominant single wave number spectrum band is nearly centered at the dominant wave number l_0. Here, $(\mathbf{l}_0)$ is the spectrum range at which the contribution is important to ϕ; outside of $(\mathbf{l}_0)$, the contribution is negligible.

Therefore, in this case, we have

$$\phi(\mathbf{r},t) = \int\int\int \Psi_{m,n,k} e^{i(\mathbf{l}\cdot\mathbf{r}-\sigma t)}\,dm\,dn\,dk$$

$$= \int\int\int \Psi_{m,n,k} e^{i[(\mathbf{l}-\mathbf{l}_0)\cdot\mathbf{r}-(\sigma-\sigma_0)t]} e^{i(\mathbf{l}_0\cdot\mathbf{r}-\sigma_0 t)}\,dm\,dn\,dk, \tag{2.11}$$

where $e^{i(\mathbf{l}_0\cdot\mathbf{r}-\sigma_0)}$ is independent of the integration. Hence, eq. (2.11) can be written as the following *single wave packet* form:

$$\phi(\mathbf{r},t) = A(\mathbf{r},t) e^{i(\mathbf{l}_0\cdot\mathbf{r}-\sigma_0 t)}, \tag{2.12}$$

where

$$A(\mathbf{r},t) = \int\int\int \Psi_{m,n,k} e^{\Delta\mathbf{l}\cdot[\mathbf{r}-(\frac{\Delta\sigma}{\Delta\mathbf{l}})t]}\,dm\,dn\,dk \tag{2.13}$$

and

$$\Delta\mathbf{l} = \mathbf{l} - \mathbf{l}_0. \tag{2.14}$$

The wave packet can be approximated in this case as

$$A(\mathbf{r},t) \simeq \int\int\int \Psi_{m,n,k} e^{i\Delta\mathbf{l}\cdot[\mathbf{r}-(\frac{\partial\sigma}{\partial\mathbf{l}})_0 t]}\,dm\,dn\,dk, \tag{2.15}$$

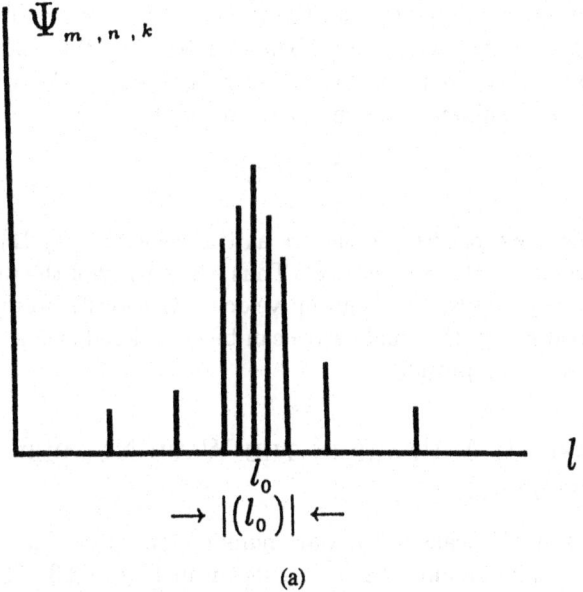

(a)

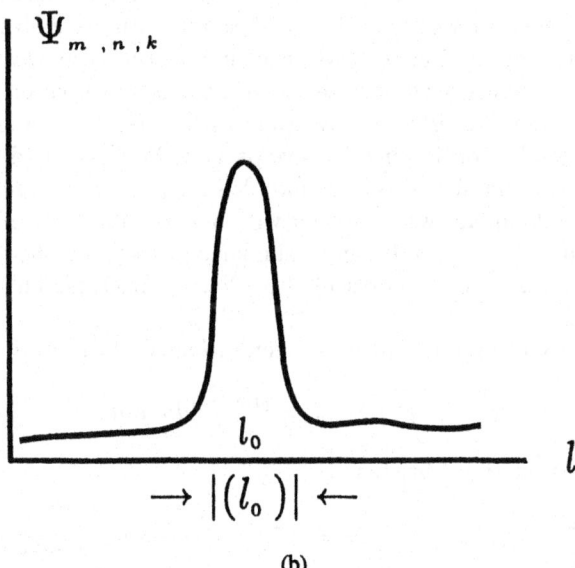

(b)

FIGURE 2.2. Typical spectrum distribution for a single wave packet. (a) Discrete wave spectrum and (b) continuous spectrum.

where integration is only taken over the whole space of wave number spectrum of $(\mathbf{1}_0)$, and $\Delta\sigma/\Delta\mathbf{1}_0$ is approximated by $(\partial\sigma/\partial\mathbf{1})_0$, which is the differentiation of wave frequency σ with respect to the total wave number, evaluated at the dominant wave number, provided

$$\left|\frac{\Delta\mathbf{1}}{\mathbf{1}_0}\right| \ll 1.$$

Therefore, the wave packet propagates at the velocity $(\partial\sigma/\partial\mathbf{1})$, which is the group velocity of the wave packet. Thus, the energy of the disturbance system is propagating with this group velocity. Obviously, from the argument developed so far, the single wave can be considered as a special wave packet of constant amplitude.

2.2.3 CASE II: N PEAKS IN THE WAVE NUMBER SPECTRUM

Suppose there are N peaks in the wave number spectrum, that is, N dominant waves or wave number bands, as shown in Figure 2.3. The N dominant wave number spectrum bands are centered at the dominant waves with wave numbers l_j $(j = 1, \ldots, N)$, respectively. Figure 2.3a is a typical discrete wave spectrum distribution for this case. Figure 2.3b is a typical distribution for the continuous wave spectrum. In this case, we cannot represent the disturbance system in terms of a single wave packet, as we did in the above case. Nevertheless, we can easily extend the above idea to the present case by considering N wave packets. We first divide the whole range of the wave number spectrum into N subranges. Over each subrange there occurs a dominant wave number with a wave number spectrum band (l_j), as shown in Figure 2.3. It can be similarly proven that the disturbance system in this case can be represented by N wave packets. The arguments are as follows:

The disturbance system can be written in Fourier form, as in (2.8),

$$\phi = \int\int\int \Psi_{m,n,k}e^{i(\mathbf{1}\cdot\mathbf{r}-\sigma t)}dmdndk. \tag{2.16}$$

Accordingly, it can be rewritten

$$\phi = \sum_{j=1}^{N}\int\int\int \Psi_{m,n,k}e^{i\Delta\mathbf{1}_j[\mathbf{r}-(\frac{\Delta\sigma}{\Delta\mathbf{1}_j})t]}e^{i(\mathbf{1}_j\cdot\mathbf{r}-\sigma_j t)}dmdndk, \tag{2.17}$$

where the integration is taken over the jth subrange of the wave number spectrum space, respectively. By using the same kind of approximation as in the previous case, we obtain

$$\phi \simeq \sum_{j=1}^{N}A_j(\mathbf{r},t)e^{i(\mathbf{1}_j\cdot\mathbf{r}-\sigma_j t)}, \tag{2.18}$$

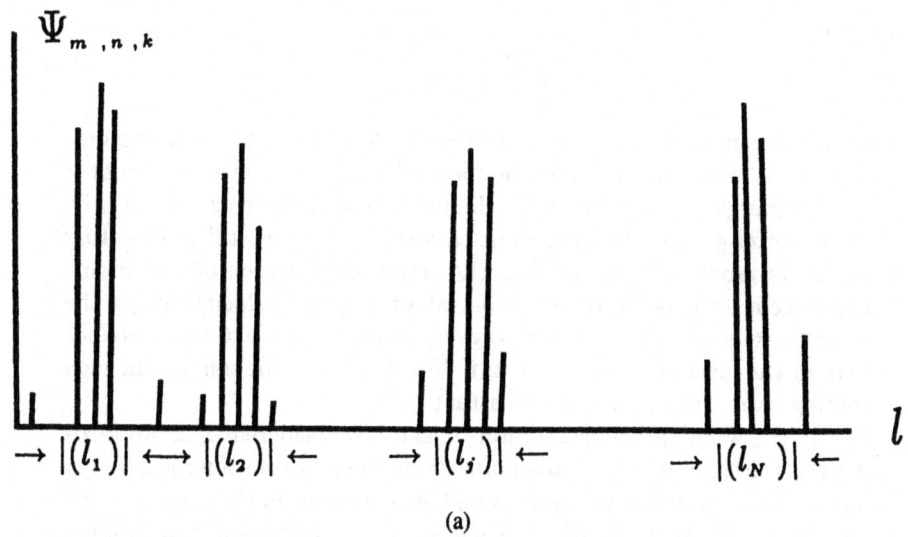

(a)

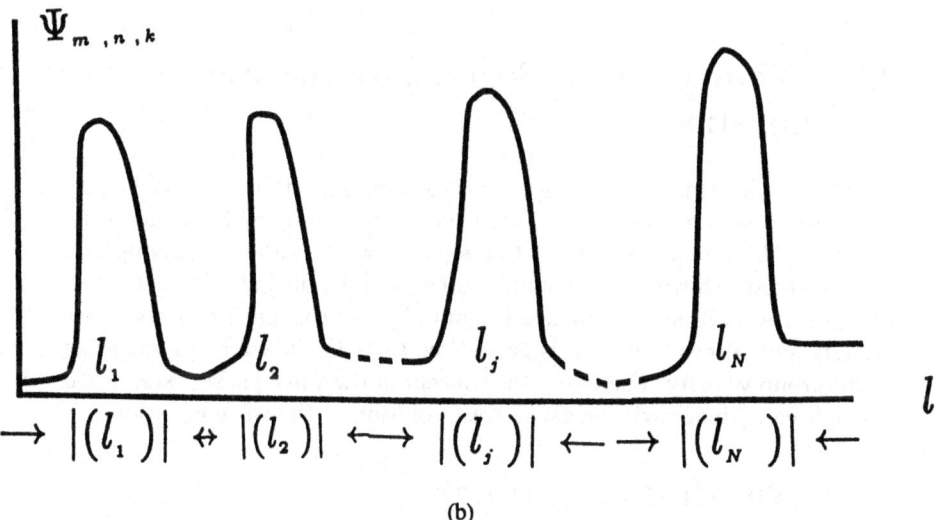

(b)

FIGURE 2.3. Typical spectrum distribution for N wave packets. (a) Discrete wave spectrum and (b) continuous spectrum.

where

$$A_j(\mathbf{r}, t) = \int\int\int \Psi_{m,n,k} e^{i\Delta 1_j[\mathbf{r} - (\frac{\partial\sigma}{\partial 1})_j t]} \, dm \, dn \, dk \qquad (2.19)$$

and

$$\Delta 1_j = 1 - 1_j, \qquad j = 1, \ldots, N, \qquad (2.20)$$

provided

$$\left|\frac{\Delta \mathbf{1}_j}{\mathbf{1}_j}\right| \ll 1.$$

Thus, the disturbance system is in the form of N *wave packets*. The integration in (2.18) has taken over only the space of the wave number spectrum of $(\mathbf{1}_j)$. The energy propagation of the disturbance system depends on all the N wave packet group velocities, while each wave packet is still propagating along at its group velocity. In fact, the actual propagation of the disturbance system depends on the extreme values of group velocity among the N wave packets. In general, it is always possible to represent a disturbance system in the form of a wave packet, if N is allowed to be infinity. In such a limiting case, the amplitude is constant.

Hence, the wave packet representation can be considered as an approximation of waves, in which, instead of a single wave, we consider a group of waves within a subrange of a wave number spectrum, in the case of single wave packet, and N groups of waves within N subranges of a wave number spectrum, in the case of a N wave packet.

2.3 The Asymptotic Solution for the Wave Equation

We have so far considered the wave packet representation for an arbitrary disturbance system and found that each wave packet will propagate its energy at the group velocity. In this section, we consider the asymptotic behavior of an arbitrary disturbance system, that is, the behavior as $t \to \infty$. Moreover, we will show that a wave initially is propagating at its phase velocity and, after a sufficient lapse of time t ($\gg P$), it will be propagating at the group velocity. Therefore, the concept of the wave packet established above is consistent with the asymptotic solution of the wave equation.

2.3.1 THE GENERAL SOLUTION

As stated earlier, we can obtain the general solution by superposing each component of a Fourier expansion, that is,

$$\phi(x,t) = \int_{-\infty}^{+\infty} A(m)e^{i(mx-\sigma t)}dm, \qquad (2.21)$$

in a one-dimensional problem, where the frequency σ is the function of the wave number and the physical parameter, that is, $\sigma = \sigma(m, \lambda)$, is determined by the dispersion relation. The spectrum function $A(m)$ takes care of the initial condition. In principle, it is always possible to construct the spectrum function in a given problem by this standard method. The solution (2.21) corresponds to the initial condition by

$$\phi(x,0) = \int_{-\infty}^{+\infty} A(m)e^{imx}dm, \qquad (2.22)$$

which is the Fourier integral for $\phi(x,0)$ and, consequently, given $\phi(x,0)$, $A(m)$ can be evaluated.

As far as the asymptotic behavior is concerned, we are now only interested in determining how the general solution (2.21) behaves after a large time lapse; for instance, when $t \gg t_c$, where t_c is some characteristic time, like period P, associated with the wave. There is a method, called the *steepest descent method* or *the saddle-point method*, to obtain this asymptotic solution. This method is the simplest one available and is included in many books, for example, Jeffreys and Jeffreys (1956) or Dennery and Krizywicki (1967).

The solution (2.21) can be written in the following form:

$$\phi(x,t) = \int_{-\infty}^{+\infty} A(m)e^{it\chi(m)}dm, \qquad (2.23)$$

where the phase function $\chi(m)$ is given by

$$\chi(m) = \frac{x}{t}m - \sigma(m). \qquad (2.24)$$

We assume that $\chi(m)$ is analytic in the complex m-plane for a fixed value of x/t. This assumption is always valid in most physical problems of interest.

The saddle-point theory says that the asymptotic solution of (2.21), as $t \to \infty$, can be reached when the phase function $\chi(m)$ attains an extreme value, that is, a stationary value. Accordingly, the *saddle point* is defined as a point at which the derivative of the phase function $\chi(m)$, with respect to wave number m, is zero. Therefore, in the present case, the saddle points are given by

$$\left(\frac{\partial \chi(m)}{\partial m}\right)\bigg|_{(x/t)\ \text{fixed}} = 0, \qquad (2.25)$$

that is,

$$\frac{x}{t} = \frac{\partial \sigma(m)}{\partial m}, \qquad (2.26)$$

provided

$$\frac{\partial^2 \sigma}{\partial m^2} \neq 0.$$

On solving for m, we obtain the saddle points

$$m_j = m_j(x/t). \qquad (2.27)$$

Corresponding to the saddle-point m_j, the asymptotic solution by this method gives the following form for the general solution (2.21):

$$\phi \simeq \int_{-\infty}^{+\infty} A(m) e^{it[\chi(m_j) + \frac{1}{2}(\frac{\partial \chi^2(m_j)}{\partial m^2})(m - m_j)^2]}, \qquad (2.28)$$

$$= \sqrt{\frac{2\pi}{t|\sigma''(m_j)|}} A(m_j) e^{i[t\chi(m_j) + \alpha]}, \qquad (2.29a)$$

$$= \sqrt{\frac{2\pi}{t|\sigma''(m_j)|}} A(m_j) e^{i[m_j x - \sigma(m_j)t] + i\alpha}, \qquad (2.29b)$$

as $t \to \infty$, where

$$\alpha = \frac{\pi}{4}, \qquad (2.30a)$$

if $\sigma''(m_j) < 0$, that is, if χ has a minimum value at m_j;

$$\alpha = -\frac{\pi}{4}, \qquad (2.30b)$$

if $\sigma''(m_j) > 0$, that is, if χ has a maximum value at m_j.

If there is more than one saddle point, say, J saddle points, taking all saddle points into account, we have

$$\phi(x,t) \simeq \sum_{j=1}^{J} \sqrt{\frac{2\pi}{t|\sigma''(m_j)|}} A(m_j) e^{i[m_j x - \sigma(m_j)t] - i\frac{\pi}{4} \text{sgn}[\sigma''(m_j)]}, \qquad (2.31)$$

where

$$\text{sgn}\,\sigma''(m_j) = +1, \text{ if } \sigma''(m_j) > 0, \qquad (2.32a)$$
$$= -1, \text{ if } \sigma''(m_j) < 0. \qquad (2.32b)$$

From the results (2.29b), the following characteristics of the asymptotic solution are true:

1. Asymptotically, the wave is propagated along with the group velocity, as shown by eq. (2.26);

2. it represents a locally harmonic wave that is not uniform in the sense that m_j and $\sigma(m_j)$ vary with x/t, in spite of the fact that the initial state of the wave was not harmonic;

3. ultimately, there is a phase difference, $\pi/4$ or $-\pi/4$;

4. the amplitude of wave, given by

$$\sqrt{\frac{2\pi}{t|\sigma''(m_j)|}}A(m_j),\qquad(2.33)$$

decreases inversely as the square root of t over distances and times of the order of x and t themselves.

One might be surprised at the first sight of the decay of the amplitude of the wave in a nondissipative system. In fact, it can be quite clear that this is an outcome of the distribution of the energy of the initial wave over the increasingly long wave train or wave packet that results from an ever-increasing dispersion with time. For instance, the energy between wave numbers m_j and $m_j + dm$ is initially proportional to $A^2(m_j)dm$. After time t, the distance between these two wave numbers becomes

$$d = |t\sigma'(m_j) - t\sigma'(m_j + dm)| \simeq t|\sigma''(m_j)|dm,\qquad(2.34)$$

so that now the energy density is proportional to

$$\frac{A^2(m_j)dm}{t|\sigma''(m_j)|dm}.\qquad(2.35)$$

Since the energy density of the wave is proportional to the square of the amplitude, the amplitude of the wave at t is proportional to

$$\frac{A(m_j)}{\sqrt{t|\sigma''(m_j)|}}.\qquad(2.36)$$

When $\sigma''(m_j) = 0$, using the theory of the asymptotic behavior of Fourier integrals, Lighthill (1965) has shown that the asymptotic solution due to the saddle point m_j can be written as

$$\phi(x,t) \simeq \frac{A(m_j)\left(\frac{1}{3}\right)!\sqrt{3}}{\left[\frac{1}{6}t|\sigma'''(m_j)|\right]^{1/3}}e^{i[m_j x - \sigma(m_j)t]},\qquad(2.37)$$

provided

$$\sigma'''(m_j) \equiv \frac{\partial^3 \sigma}{\partial m^3} \neq 0.$$

Therefore, the amplitude of the wave now decreases as the inverse cubic root of time t.

2.3.2   PROPAGATION OF ENERGY

Let us now determine the velocity with which the energy is propagated in a dispersive wave. Suppose there are two waves with wave numbers m_1

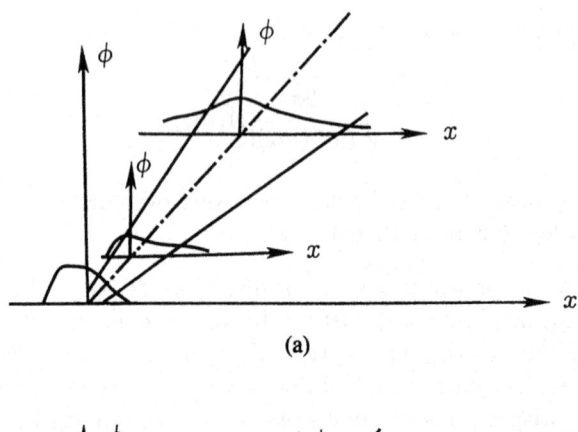

(a)

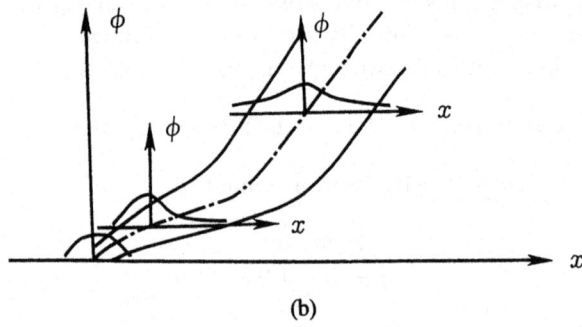

(b)

FIGURE 2.4. Dispersion of waves at $x = 0$, $t = 0$. (a) Conserved waves and (b) general waves.

and m_2 starting from the point $x = 0$, at $t = 0$, with velocities c_1 and c_2, respectively, as shown in Figure 2.4. After a sufficient time, these waves will be at x_1 and x_2, where

$$x_1 = c_{g1}(m_1')t \qquad (2.38a)$$

and

$$x_2 = c_{g2}(m_2')t, \qquad (2.38b)$$

provided c_{g1} and c_{g2} are the group velocities corresponding to wave numbers m_1' and m_2', respectively. At time t, the value of ϕ is given by (2.29b), which approximates the harmonic wave if we neglect the slow variations in m_j and $\sigma(m_j)$ with space and time. Therefore, the energy E of the wave between x_1 and x_2 may be taken as proportional to

$$\int_{c_{g1}}^{c_{g2}} \frac{2\pi A^2 \sin^2[m_j x - \sigma(m_j)t + (\pi/4)\mathrm{sgn}(\sigma''(m_j))]dx}{t|\sigma''(m_j)|}. \qquad (2.39)$$

If we consider values of t and $P \ll t \ll \tau$, then there will be several waves between x_1 and x_2, and we treat m_j and $\sigma(m_j)$ as approximately constant. Therefore,

$$E \propto \int_{c_{g1}}^{c_{g2}} \frac{\pi A^2(m_j)dx}{t|\sigma''(m_j)|}, \tag{2.40}$$

which, on substituting $x = c_g t$, reduces to

$$E \propto \int_{c_{g1}}^{c_{g2}} \frac{\pi A^2(m_j)dc_g}{\sigma''(m_j)}. \tag{2.41}$$

However, we have

$$m_j = m_j(c_g), \quad c_g = \sigma'(m_j) \tag{2.42}$$

from (2.26). Thus, we can conclude that the energy between the two points of the wave starting at the origin is time independent, provided the wave numbers are conserved. In this case, the lines along which the energy is propagated are straight lines, as shown in Figure 2.4a. Nevertheless, if the wave numbers are not conserved, then the lines along which the energy is propagated will not be straight, as shown in Figure 2.4b. These lines are called *characteristic lines of propagation of energy*.

From the above discussion, we can make the following interesting and important point. If we wish to follow a wave of a given wave number, we have to move at the group velocity c_g. Thus, after time t ($\gg P$), this wave will be at

$$x_1 = c_g t, \tag{2.43}$$

not at

$$x_2 = ct, \tag{2.44}$$

where c_g and c are the group velocity and the phase velocity discussed above. If the group velocity is varying with space and time, then this wave will be at

$$x_1 = \int_0^t c_g dt. \tag{2.45}$$

Therefore, it is important to keep in mind that a wave is initially propagated at the phase velocity and, after the lapse of a sufficiently long time, it is propagated at group velocity.

Figure 2.5 shows the switching of a wave moving at phase velocity to group velocity. Figure 2.5a shows a case in which the wave number is conserved, as is the group velocity. Figure 2.5b shows a case in which the wave number is not conserved, nor is the group velocity. The question under what conditions and how the wave number changes along these characteristic lines will be addressed in Section 2.5.

So far, we have shown that, essentially, the concept of wave packet representation for an arbitrary disturbance system is consistent with the asymptotic behavior of the wave equation.

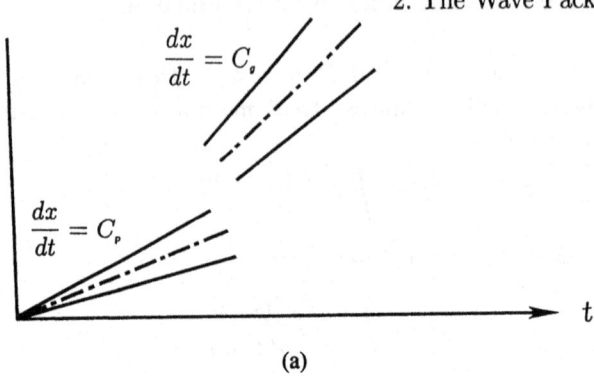

(a)

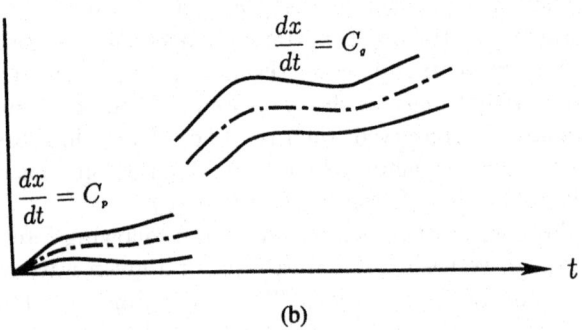

(b)

FIGURE 2.5. Switching of wave propagation with phase velocity to group velocity. (a) Conserved waves and (b) general waves.

2.4   The WKB Method

The WKB method is used to solve a differential equation approximately. The exact solution of the equation may be some unknown function of overwhelming complexity. By use of the WKB approximation, order by order, however, we might obtain the approximate solution, which consists of exponentials of elementary integrals of algebraic functions, and well-known special functions, such as the Airy function or the parabolic cylinder function. The WKB approximation is suitable for linear differential equations of any order, for initial-value and boundary-value problems, and for eigenvalue problems. It may also be used to evaluate the integrals of the solution of a differential equation.

2.4.1 THE WKB APPROXIMATION

In general wave equations, dissipative and/or dispersive waves are common
phenomena and are both characterized by exponential behavior, when the
exponential is real in the dissipative wave and imaginary in the dispersive
one. Therefore, for a wave equation (in general, a differential equation)
that exhibits either or both kinds of behavior, it is natural to seek an
approximation solution in the following form:

$$\phi(x,t) \sim A(X,T)e^{i\theta(X,T)/\varepsilon}, \qquad \varepsilon \to 0_+. \tag{2.46}$$

The phase $\theta(X,T)$ is assumed nonconstant, and $A(X,T)$ and $\theta(X,T)$ are
slowly varying functions in which the variables are introduced as follows:

$$X = \varepsilon x, \qquad T = \varepsilon t, \tag{2.47}$$

where ε is a small parameter. The exponential approximation in (2.46)
is conventionally called the *WKB approximation,* named after Wentzel,
Kramers, and Brillouin, who popularized the theory. Keller (1958) modified
the method and proposed its present form.

2.4.2 FORMAL EXPANSION AND VALIDITY CONDITION

The exponential approximation in (2.46) also can be represented by ex-
panding A and θ functions in a series of powers of ε. Therefore, we can use
the following more general form:

$$\phi(x) \sim e^{\left(\frac{1}{\varepsilon}\sum_{n=0}^{\infty}\varepsilon^n\Theta_n\right)}, \qquad \varepsilon \to 0_+ \tag{2.48}$$

for a one-dimensional problem. This is the *formal WKB expansion.* All
WKB approximations are derived from this form.

In order that the WKB approximation (2.48) be valid on an interval, it
is necessary that the series $\sum \varepsilon^{n-1}\Theta_n$ be an asymptotic series in ε as $\varepsilon \to 0_+$ uniformly, for all x on the interval. This requires that the asymptotic
relations

$$\begin{cases} \Theta_1 \ll \dfrac{1}{\varepsilon}\Theta_0, \ \varepsilon \to 0 \\ \varepsilon\Theta_2 \ll \Theta_1, \ \varepsilon \to 0 \\ \quad \cdots \\ \varepsilon^n\Theta_{n+1} \ll \varepsilon^{n-1}\Theta_n, \ \varepsilon \to 0 \end{cases} \tag{2.49}$$

hold uniformly in x. For the WKB series truncated at the term $\varepsilon^{N-1}\Theta_N$
to be a good approximation to ϕ, the next term must be small compared
with 1 for all x in the interval of approximation:

$$\varepsilon^N\Theta_{N+1} \ll 1, \qquad \varepsilon \to 0. \tag{2.50}$$

If this relation hold, then

$$e^{(\varepsilon^N\Theta_{N+1})} = 1 + O(\varepsilon^N\Theta_{N+1}), \qquad \varepsilon \to 0. \tag{2.51}$$

Thus, the relative error between ϕ and the WKB approximation is small, that is,

$$\frac{\phi - e^{(1/\varepsilon \sum_{n=0}^{N} \varepsilon^n \Theta_n)}}{\phi} \to \varepsilon^N \Theta_{N+1}, \qquad \varepsilon \to 0. \qquad (2.52)$$

For the validity of the WKB approximation, both (2.50) and (2.52) must be satisfied.

Finally, it should be pointed out that in the literature the name *approximation of geometrical optics* can be found for a case in which we only retain the first term in the formal WKB expansion. This is the approximation

$$\phi(x) \to e^{\Theta_0/\varepsilon}. \qquad (2.53)$$

Furthermore, if the first two terms are retained, the approximation is called the *approximation of physical optics*, that is,

$$\phi(x) \to e^{\Theta_0/\varepsilon + \Theta_1}. \qquad (2.54)$$

The relative error between ϕ and the approximation of physical optics is of order $\varepsilon \Theta_2$, which vanishes uniformly with ε if Θ_2 is bounded. In general, the approximation of physical optics expresses the leading asymptotic behavior of ϕ, whereas the approximation of geometrical optics contains just the controlling factor of the leading behavior.

Compared with our previous discussion, it is found that the wave packet representation in the single wave packet form and the WKB approximation are consistent with each other if the fact the slowly varying variables are used in the WKB approximation and fast variables have been used in the wave packet form (2.12) is taken into account. Their variables are related by the relation (2.47). Hence, the form that will be used the most in the rest of the chapters is the form (2.46), which agrees with the wave packet approximation. We simply term them a wave packet or a WKB approximation.

2.5   General Theory

We earlier discussed the representation of disturbance system in the form of the wave packet and found that the energy of the wave packet is propagated along the characteristic lines determined by the group velocity. The question raised in Section 2.3 about the change of wave number of the wave packet is answered in this section.

Some properties of the wave packet are discussed. These properties also hold for general linear and nonlinear waves. We will begin with an example of a general wave equation in geophysical fluid dynamics with a time-independent, non-homogeneous medium. Then we study the general theory of wave packet, including the eikonal equation, the laws of the wave number, and frequency conservation. Wave-action conservation is obtained by using the Hamilton principle, that is, a variational approach. The energy exchange between wave packet and basic flow is also considered at the end of this section.

2.5.1 AN EXAMPLE: THE ROSSBY WAVE PACKET

Suppose there is a two-dimensional problem, which can be described by the following wave equation:

$$\frac{\partial}{\partial t}\nabla^2\phi + \beta\frac{\partial\phi}{\partial x} = 0. \tag{2.55}$$

This equation is a typical wave equation in geophysical fluid dynamics and describes the Rossby waves. Instead of taking β constant, we allow β to change with space. That is, we take β as the first derivative of the Coriolis parameter, as mentioned in Section 1.3. For large-scale motion, β is varying slowly with latitude. We use the WKB approximation (2.46), that is, the wave packet form

$$\phi \sim \Psi e^{i\theta/\varepsilon}, \tag{2.56}$$

where ε is a small parameter, which can be chosen as the ratio of the wave length of the wave packet and the wave length of the medium. Here, Ψ and θ are functions of the slowly varying variables X, Y, and T, as in (2.47), and are defined by

$$X = \varepsilon x, \qquad Y = \varepsilon y, \qquad T = \varepsilon t. \tag{2.57}$$

And Ψ can be expanded in terms of the ε series as

$$\Psi(X,Y,T) = \Psi_0(X,Y,T) + \varepsilon\Psi_1(X,Y,T) + \varepsilon^2\Psi_2(X,Y,T) + \cdots. \tag{2.58}$$

Define

$$\sigma = -\frac{\partial\theta}{\partial T}, \tag{2.59}$$

where σ is the local frequency of the wave packet,

$$m = \frac{\partial\theta}{\partial X}, \tag{2.60a}$$

and

$$n = \frac{\partial\theta}{\partial Y}, \tag{2.60b}$$

where m and n are the local wave numbers in the X- and Y-direction, respectively. From this definition, we have the relations

$$\frac{\partial\sigma}{\partial x} = -\frac{\partial m}{\partial T}, \tag{2.61a}$$

$$\frac{\partial\sigma}{\partial Y} = -\frac{\partial n}{\partial T}, \tag{2.61b}$$

and

$$\frac{\partial m}{\partial Y} = \frac{\partial n}{\partial X}. \tag{2.62}$$

Substituting the expression of wave packet, that is, (2.56), into the Rossby wave equation (2.55) and using the above relations at the lowest order, we obtain the following dispersion relation:

$$\sigma = -\frac{\beta m}{m^2 + n^2}. \tag{2.63}$$

The local frequency is the nonlinear function of the local number and the linear function of β, that is,

$$\sigma = \sigma(\beta, m, n). \tag{2.64}$$

This is obviously a dispersive wave packet, and precisely, the dispersive Rossby wave packet.

From (2.64), we obtain

$$\frac{\partial \sigma}{\partial X} = \frac{\partial \sigma}{\partial m}\frac{\partial m}{\partial X} + \frac{\partial \sigma}{\partial n}\frac{\partial n}{\partial X} + \frac{\partial \sigma}{\partial \beta}\frac{\partial \beta}{\partial X}. \tag{2.65}$$

Using (2.61) and (2.62), we can rewrite the above relation as

$$\frac{\partial m}{\partial T} + \frac{\partial \sigma}{\partial x}\frac{\partial m}{\partial X} + \frac{\partial \sigma}{\partial m}\frac{\partial m}{\partial Y} = -\frac{\partial \sigma}{\partial \beta}\frac{\partial \beta}{\partial X}. \tag{2.66}$$

Since the group velocity of this wave packet is defined by

$$C_{gX} = \frac{\partial \sigma}{\partial m} = -\frac{\beta(n^2 - m^2)}{(m^2 + n^2)^2}, \tag{2.67a}$$

$$C_{gY} = \frac{\partial \sigma}{\partial n} = \frac{2\beta mn}{(m^2 + n^2)^2}. \tag{2.67b}$$

Therefore, eq. (2.66) can be further written as

$$\frac{\partial m}{\partial T} + C_{gX}\frac{\partial m}{\partial X} + C_{gY}\frac{\partial m}{\partial Y} = -\frac{m}{m^2 + n^2}\frac{\partial \beta}{\partial X} = 0, \tag{2.68}$$

since

$$\beta = \frac{2\Omega \cos \phi}{a} \tag{2.69}$$

is independent of X.

Similarly, differentiating σ with respect to Y and using the relations (2.61b) and (2.62), we obtain

$$\frac{\partial n}{\partial T} + C_{gX}\frac{\partial n}{\partial X} + C_{gY}\frac{\partial n}{\partial Y} = -\frac{m}{m^2 + n^2}\frac{\partial \beta}{\partial Y} = \frac{\delta m}{m^2 + n^2}, \tag{2.70}$$

where

$$\delta = -\frac{\partial \beta}{\partial Y} \tag{2.71}$$

has been used and (2.70) has a positive value in the Northern Hemisphere and is similar to the previous definition (1.25).

Thus, we have obtained the equations governing the local wave numbers m and n, (2.68) and (2.70) with (2.67a,b). Similarly, by differentiating σ with respect to T, we can easily derive the equation governing σ:

$$\frac{\partial \sigma}{\partial T} + C_{gX}\frac{\partial \sigma}{\partial X} + C_{gY}\frac{\partial \sigma}{\partial Y} = 0, \tag{2.72}$$

with the aid of (2.61a) and the dispersion relation (2.63).

Now we have obtained a set of equations governing the whole evolution of the Rossby wave packet, including energy propagation and structural change. They are (2.72), (2.68), (2.70), and (2.67a,b), that is,

$$\frac{\partial \sigma}{\partial T} + C_{gX}\frac{\partial \sigma}{\partial X} + C_{gY}\frac{\partial \sigma}{\partial Y} = 0, \tag{2.73}$$

$$\frac{\partial m}{\partial T} + C_{gX}\frac{\partial m}{\partial X} + C_{gY}\frac{\partial m}{\partial Y} = -\frac{m}{m^2+n^2}\frac{\partial \beta}{\partial X} = 0, \tag{2.74a}$$

$$\frac{\partial n}{\partial T} + C_{gX}\frac{\partial n}{\partial X} + C_{gY}\frac{\partial n}{Y} = \frac{\delta m}{m^2+n^2}, \tag{2.74b}$$

$$C_{gX} = \frac{\partial \sigma}{\partial m} = \frac{\beta(n^2-m^2)}{(m^2+n^2)^2}, \tag{2.75a}$$

$$C_{gY} = \frac{\partial \sigma}{\partial n} = \frac{2\beta mn}{(m^2+n^2)^2}. \tag{2.75b}$$

From the theory of partial differential equations (e.g., Courant and Hilbert, 1963), and by the use of the characteristic method, the partial differential equations (2.73) to (2.75b) can be written as follows:

$$\frac{d\sigma}{dT} = 0, \tag{2.76}$$

$$\frac{dm}{dT} = 0, \tag{2.77a}$$

$$\frac{dn}{dT} = \frac{\delta m}{m^2+n^2}, \tag{2.77b}$$

and

$$\frac{dX}{dT} = C_{gX} = \frac{\beta(n^2-m^2)}{(m^2+n^2)^2}, \tag{2.78a}$$

$$\frac{dY}{dT} = C_{gY} = \frac{2\beta mn}{(m^2+n^2)^2}. \tag{2.78b}$$

Obviously, the characteristic direction is the direction of group velocity of the wave packet. Therefore, the results show that the problem in question has the following properties:

(i) The characteristic direction coincides with the group velocity, that is, the wave packet propagation direction.

(ii) Along the characteristic direction, the local frequency and local wave number in X-direction are both conserved.

(iii) Along the characteristic lines, the change in the local number in the Y-direction is proportional to δ.

(iv) In the β-plane approximation, that is, where δ is zero, the local wave numbers m and n are both conserved.

(v) The characteristic lines are straight lines on the β-plane.

(vi) On the δ-surface, that is, since δ is constant, the equations governing the wave number and frequency can be decoupled from the whole set of equation system. That is, they are explicitly independent of the path. We discuss the effect of δ upon the Rossby wave packet in Chapter 3.

The method developed here is sometimes called the *ray method* or the *kinematic method,* since its use has always been emphasized in the application in the propagation properties of waves, and the wave packet is propagated along the group velocity. The property (vi) is of great importance, and useful; it has not been considered previously in the literature. This special property of the system allows us to simplify the dynamics system considerably when we are primarily interested in structural changes in the wave packet. The idea is easily extended into problems in three dimensions or N-dimensions.

2.5.2 AN IMPORTANT RELATION: THE EIKONAL EQUATION

In general, the governing equation in N-dimensions can be written as

$$\mathrm{L}\phi = 0, \tag{2.79}$$

where L is an operator with parameters B_j $(j = 1, \ldots, J)$, J is the number of parameters, and B_j are assumed to slowly vary with space and time.

By the use of the WKB approximation, that is, the wave packet form,

$$\phi \sim \Psi e^{i\theta/\varepsilon}, \tag{2.80}$$

where ε is a small parameter, as before.

Introduce the slowly varying variables

$$T = \varepsilon t, \qquad X_i = \varepsilon x_i, \qquad i = 1, 2, \ldots, N, \tag{2.81}$$

and define

$$\mathbf{K} = \nabla \theta \tag{2.82a}$$

and

$$\sigma = -\frac{\partial \theta}{\partial T}, \tag{2.82b}$$

where $\mathbf{K}$ is the local wave vector and σ is the local frequency of the wave packet. From these definitions, we immediately find an important relationship,

$$\frac{\partial \mathbf{K}}{\partial T} + \nabla \sigma = 0. \tag{2.83}$$

This relationship is called the *eikonal equation*. It also can be called the *conservation of crests equation*. If we consider the vector $\mathbf{K}$ as a directional density of crests in space, and the frequency σ as a flux of crests past a fixed point, then under the slowly varying condition, crests are neither destroyed nor created, and the crest density and flux are related through the eikonal equation (2.83).

2.5.3 THEORY IN MEDIA AT REST

In media at rest, by substituting (2.80) into the equation (2.79), at the lowest order, we obtain the dispersion relation, that is,

$$\sigma = \sigma(\mathbf{K}, B_j). \tag{2.84}$$

Hence, σ is varied not only through $\mathbf{K}(\mathbf{X}, T)$, but also for constant $\mathbf{K}$, through variations in the properties of the medium itself, as expressed through parameters B_j. Substituting the dispersion relation (2.84) in the eikonal equation (2.83), we obtain

$$\frac{\partial k_i}{\partial T} + \frac{\partial \sigma}{\partial k_j} \frac{\partial k_j}{\partial X_i} + \frac{\partial \sigma}{\partial B_j} \frac{\partial B_j}{\partial X_i} = 0, \qquad i = 1, 2, \ldots, N, \tag{2.85}$$

where the convention summation has been used, and k_i is the ith component of the wave vector $\mathbf{K}$ in the X_i-direction. From the definition of $\mathbf{K}$, we have

$$\nabla \times \mathbf{K} = 0 \tag{2.86a}$$

or

$$\frac{\partial k_j}{\partial X_i} = \frac{\partial k_i}{\partial X_j}. \tag{2.86b}$$

Moreover, if we define the group velocity $\mathbf{C}_g$ by the relation

$$C_{gi} = \frac{\partial \sigma}{\partial k_i}, \tag{2.87}$$

then eq. (2.85) can be written as

$$\frac{\partial k_i}{\partial T} + \mathbf{C}_g \cdot \nabla k_i = -\frac{\partial \sigma}{\partial B_j} \frac{\partial B_j}{\partial X_i}, \qquad i = 1, 2, \ldots, N. \tag{2.88}$$

Similarly, differentiating the dispersion relation (2.84) with respect to T, and using the eikonal equation (2.83), we obtain

$$\frac{\partial \sigma}{\partial T} + \mathbf{C}_g \cdot \nabla \sigma = \frac{\partial \sigma}{\partial B_j} \frac{\partial B_j}{\partial T}. \tag{2.89}$$

According to the theory of partial differential equations, we find that eqs. (2.88) and (2.89) have the same characteristic curves, which are obtained by integrating the relationships

$$\frac{d\mathbf{X}}{dT} = \mathbf{C}_g \tag{2.90}$$

(see, e.g., Courant and Hilbert, 1962). These characteristic curves are called *rays* in the wave theory. This is the reason why people call it the *ray theory*, or the *ray method*. The characteristic curves are of great importance and describe the paths of the wave packet, since the wave packet is propagated along these curves, as shown in earlier sections. Integration of (2.90) gives the paths of the wave packet

$$\mathbf{X} - \int \mathbf{C}_g \, dT = \alpha, \tag{2.91}$$

where α is a constant vector, that varies from one path to another. Choosing the path as a kind of natural coordinate system for describing the wave packet, the time derivative along the characteristic curve, that is, by keeping α fixed, is then

$$\frac{D_g}{DT} = \left(\frac{d}{dT} \right)_\alpha \equiv \frac{\partial}{\partial T} + \mathbf{C}_g \cdot \nabla. \tag{2.92}$$

Therefore, the resulting dynamic system can be written as

$$\frac{D_g X_i}{DT} = \frac{\partial \sigma}{\partial k_i}, \tag{2.93a}$$

$$\frac{D_g k_i}{DT} = -\frac{\partial \sigma}{\partial B_j} \frac{\partial B_j}{\partial X_i}, \tag{2.93b}$$

and

$$\frac{D_g \sigma}{DT} = \frac{\partial \sigma}{\partial B_j} \frac{\partial B_j}{\partial T}. \tag{2.93c}$$

The relation among σ, B_j, and k_i is given by the lowest order of the wave equation considered in the problem, that is, the dispersion relation (2.84).

From the dynamic system (2.93) governing the evolution of the wave packet, we can propose the following theorems:

(i) *Equivalence to Hamilton Equations.* The first two equations of (2.93) are identical in form to Hamilton's equations of classical particle dynamics (e.g., Goldstein 1980; Arnold 1978). In our equations, σ is equivalent to the role of a Hamiltonian and k_i is equivalent to the generalized momenta in particle dynamics.

(ii) *Structure Independence.* The independence of the equations governing the structure of the wave packet. When the right-hand side terms in (2.93b) are explicitly independent of the spatial variables, the equations governing the structure of the wave packet can be decoupled from the complete dynamic system. That is, the second equation (2.93b) does not explicitly depend on the first equation (2.93a), though the first equation, in general, still depends on the second one. In other words, the change in the wave number of the wave packet depends only implicitly on the change in the characteristic curve. This is true in many practical problems. The structure independence theorem enables us to handle the problem easily, to simplify the problem, and moreover, to allow us to look, with more insight, at the problem, as is shown in the following chapters. This result has not been discussed in the literature.

(iii) *The frequency conservation.* The result (2.93c) states that if the medium is time independent, then the frequency along the characteristic curve is conserved. Equation (2.93c) is the frequency conservation equation. The frequency along the characteristic curve changes only when the medium changes with time.

(iv) *The wave number conservation.* If the medium is uniform, that is, homogeneous, then eq. (2.93b) tells us that local wave numbers along the characteristic curve are conserved. Furthermore, the change in the local wave number in the X_i direction can only be the result of an inhomogeneity of the medium in that direction.

(iv) *Straight characteristic curves.* If the medium is homogeneous, then the characteristic curves will be straight curves for nondispersive waves. In general, characteristic curves are not straight.

(v) In general, any variation of the wave number in the x_i direction along the characteristic curve must be the result of a spatial inhomogeneity in that direction. Any variation of the frequency along the characteristic curve must be the result of a temporal inhomogeneity.

Furthermore, in a completely homogeneous and time-independent medium, both the frequency and the wave number are conserved along the characteristic curves.

2.5.4 THEORY IN MOVING MEDIA

The preceeding results can be easily generalized for the case in moving media. Let $\mathbf{U}(\mathbf{X}, T)$ be a basic flow or a mean flow, that is, the ambient current, which is varied over length and time scales well in excess of the wave length and the period of the wave packet. Since the wave packet is advected by the basic flow, the local frequency observed by a stationary observer becomes

$$\sigma = \mathbf{K} \cdot (\mathbf{C}_0 + \mathbf{U}), \tag{2.94}$$

where $\mathbf{C}_0$ is the phase velocity measured relative to the moving medium. To distinguish the quantities measured by a moving observer with velocity $\mathbf{U}$ from those measured in a stationary, that is, at rest, frame of reference, we use the subscript 0 to denote the quantities measured in a moving frame of reference. In other words, the frequency σ_0 denotes the frequency as measured by an observer traveling with velocity $\mathbf{U}$. This relative frequency, σ_0, is still given by the dispersion relation (2.84).

Though the translation between the two coordinates will not affect the wave number, the frequency has shifted by the amount $\mathbf{K} \cdot \mathbf{U}$, that is,

$$\sigma = \sigma_0(\mathbf{K}, B_j) + \mathbf{K} \cdot \mathbf{U}. \tag{2.95}$$

The frequency σ_0 is called the *Doppler-shifted frequency*.

Therefore, the results in media at rest can be adapted for this case. The characteristic curve is now given by

$$\frac{D_g \mathbf{X}}{DT} = \frac{\partial \sigma_0}{\partial \mathbf{K}} + \mathbf{U}. \tag{2.96}$$

And the wave number equation can be derived as

$$\frac{D_g \mathbf{K}}{DT} = -\frac{\partial \sigma_0}{\partial B_j} \nabla B_j - k_j \nabla U_j, \tag{2.97}$$

while the frequency equation will read

$$\frac{D_g \sigma}{DT} = \frac{\partial \sigma_0}{\partial B_j} \frac{\partial B_j}{\partial T} + k_j \frac{\partial U_j}{\partial T}. \tag{2.98}$$

In (2.96) to (2.98), U_j is the jth component of the ambient current $\mathbf{U}$ in the X_j-direction, and the differentiation along the characteristic curve is now expressed as

$$\frac{D_g}{DT} = \left(\frac{d}{dT}\right)_\alpha$$

$$= \frac{\partial}{\partial T} + (\mathbf{C_{g0}} + \mathbf{U}) \cdot \nabla \tag{2.99a}$$

and

$$\mathbf{X} - \int (\mathbf{C_{g0}} + \mathbf{U}) dT = \alpha. \tag{2.99b}$$

Again, α is the constant vector, which differs from one characteristic curve to another.

From eq. (2.98), it can be concluded that, in the time-independent medium, if the basic flow is steady, the frequency is conserved along the characteristic curve. Equation (2.97) shows that, in homogeneous medium, if the basic flow is uniform, the wave number is also conserved. Furthermore, we can conclude that the equations governing wave number and frequency can be decoupled from eq. (2.46), that is, the characteristic equation if and only if the medium properties B_j and the basic flow are both at their most linear function of space and time. Nevertheless, we can obtain theorems about the wave packet that are similar to those equations in media at rest, as stated in the previous section.

It should be pointed out that, in fact, if we consider the basic flow $\mathbf{U}$ as one of the medium properties B_j, both theory and conclusions in moving medium can be directly obtained from the theory in media at rest, that is, (2.93); and so can the theorems.

2.5.5   WAVE-ACTION CONSERVATION AND A VARIATIONAL APPROACH

The variational method in wave theory was introduced by Whitham (1965a,b, 1974) and further developed by Bretherton and Garrett (1968) and Garrett (1968). The entity that has been termed *wave-action* was first proposed by Whitham (1965a,b) as an essential part of a general, powerful approach to progressive waves in nonlinear conservative systems. He defined wave-action density and flux in terms of derivatives of an averaged Lagrangian L with respect to frequency and wave number, and showed that wave-action obeys a conservation law adiabatically. Bretherton and Garrett (1968) and Garrett (1968) generalized the result for moving media. Hayes (1970) extended the result and showed that wave-action is conserved in general problems both for local waves and modal waves. Progressive waves are characterized by a frequency and wave number vector in some propagation space. If the propagation space is the entire physical space $(\mathbf{x}, t)$ the waves are termed *local*. In this case, the quantity L is the local Lagrangian density averaged over phase. If the waves are not local, they are termed *modal,* and the physical space is the product of the propagation space and a cross space; in this case, the quantity L is the local Lagrangian density averaged over phase and integrated over the cross space. Modal waves are characterized by distributions in cross space corresponding to a particularly eigensolution or mode.

The advantage of a variational approach is that it puts nonlinear problems on the same ground as linear ones and allow an unprecedented degree of generality in the treatment of slowly varying waves.

The first step in this approach is to write the governing equations in the form of Euler–Lagrange. Dynamic equations such as the fluid dynamic equations, in the Euler–Lagrange, can be written as

$$\frac{\partial L}{\partial Z_i} - \frac{\partial}{\partial t}\left[\frac{\partial L}{\partial(\partial Z_i/\partial t)}\right] - \frac{\partial}{\partial x_j}\left[\frac{\partial L}{\partial(\partial Z_i/\partial x_j)}\right] = 0. \tag{2.100}$$

It should be noted that L is not uniquely defined, though it may be written in the form $L = T - V$, where T is the kinetic energy density and V is the potential energy density. The dynamic equations may be also derived from a *variational principle,* that is, *Hamilton's principle,* of the form

$$\delta \int \int \int L(Z, Z_x, Z_t; \mathbf{x}, t) dx dt = 0, \tag{2.101}$$

where Z stands for the dependent variables, for instance, $\mathbf{u}$, p, and ρ; derivatives are denoted by subscripts; and L is the Lagrangian density, defined above. Integration is taken over all space and time. The reader can find more complete information in standard textbooks of the calculus of variations and classical mechanics (e.g., Gelfand and Fomin, 1963; Arnold, 1978).

Whitman (1965a) assumed that the slow variation of a wave packet with slowly varying amplitude and with a frequency such as that discussed before, could be derived from an averaged variational principle,

$$\delta \int \int \mathrm{L}(\sigma, \mathbf{K}, A; \mathbf{x}, t) dx dt = 0, \tag{2.102}$$

where L is the time average of L over a period τ; and

$$\mathrm{L}(\sigma, \mathbf{k}, A; \mathbf{x}, t) = \frac{1}{\tau}\int_0^\tau L dt. \tag{2.103}$$

The averaging process eliminates the dependence on the short time and space scales characterizing the local variation in favor of the slow variations of wave parameters σ, $\mathbf{k}$, and A. The behavior of slowly varying properties is then given by the *Euler–Lagrange equations* (2.100) for the *averaged Lagrangian* L. Since σ and $\mathbf{k}$ are related to the phase function θ through

(2.82), we note that there are really only two dependent variables in L: θ and A or S and A. Thus,

$$L(\sigma, \mathbf{k}, A; \mathbf{x}, t) = L(S_t, \nabla S, A; \mathbf{x}, t), \tag{2.104}$$

where

$$\sigma = -\frac{\partial S}{\partial t}, \qquad \mathbf{k} = \nabla S. \tag{2.105}$$

The relation between S and θ is given by

$$S = \frac{\theta}{\varepsilon}, \tag{2.106}$$

where S is in the fast variable space and θ is in the slowly varying space.

Let $Z_1 = A$ and, replacing L by L in the Euler–Lagrange equations (2.100), then, from the averaged Hamilton's principle (2.102), we obtain

$$L_A = 0. \tag{2.107}$$

Similarly, let $Z_2 = S$, to derive

$$\frac{\partial}{\partial t} L_\sigma - \frac{\partial}{\partial x_i} L_{k_i} = 0. \tag{2.108}$$

Subscripts have been used to denote differentiation with respect to the explicit dependence of L on the dependent variables $(A, \sigma, \mathbf{k})$ and independent variables $(\mathbf{x}, t)$, where the conventional derivatives include a complete differentiation with respect to an independent variable, while others are kept fixed. For example,

$$\frac{\partial L}{\partial x_i} = L_\sigma + L_{k_j} \frac{\partial k_j}{\partial x_i} + L_A \frac{\partial A}{\partial x_i} + L_{x_i}. \tag{2.109}$$

The Euler–Lagrange equation (2.107) states a relation between σ, $\mathbf{k}$, and A, which is a generalized dispersion relation of (2.84) for nonlinear waves. This dispersion relation may be written in implicit form:

$$G[\sigma, \mathbf{k}, A; B_j(\mathbf{x}, t)] = 0. \tag{2.110}$$

For small amplitude waves on a stationary basic state, that is, media at rest, L is proportional to A^2 and is given by

$$L = G(\sigma, \mathbf{k}; B_j) A^2. \tag{2.111}$$

The averaged variational principle then gives

$$G = 0, \tag{2.112}$$

that is, the nonlinear dispersion relation, in general, so that

$$L \equiv 0, \tag{2.113}$$

which implies the *principle of the energy equipartition*, that is, on the average, that the kinetic energy is equal to the potential energy.

The group velocity $\mathbf{C}_g$ can be derived from the variational principle through the relation

$$C_{gi} = \left. \frac{\partial \sigma}{\partial k_i} \right|_{\mathbf{L}=0} = \frac{\mathbf{L}_{k_i}}{\mathbf{L}_\sigma}. \qquad (2.114)$$

Therefore, the second Euler equation (2.108) becomes

$$\frac{\partial}{\partial t}\mathbf{L}_\sigma + \frac{\partial}{\partial x_i}(C_{gi}\mathbf{L}_\sigma) = 0. \qquad (2.115)$$

The equation (2.115) states that $\mathbf{L}_\sigma$ is a conserved quantity; it is termed *wave-action*. This equation is called the *wave-action conservation equation*. The specific form of $\mathbf{L}_\sigma$ will be derived later.

In the calculus of variations, it has been proven that for any dynamic system described by a variational principle of the form (2.102), for example, there exists a conservation equation corresponding to any group of transformations with respect to which the variation principle is invariant. This is called *Noether's theorem*. The proof of the theorem can be found in the book mentioned above. Invariance with respect to time implies energy conservation. Translational invariance in space implies momentum conservation. By the use of the eikonal equation (2.83), two variational Euler equations (2.110) and (2.107), and the notation of (2.109), it can be easily verified that

$$\frac{\partial}{\partial t}(\sigma\mathbf{L}_\sigma - \mathbf{L}) - \frac{\partial}{\partial x_i}(\sigma\mathbf{L}_{k_i}) = -\mathbf{L}_t. \qquad (2.116)$$

If $\mathbf{L}$ is invariant with respect to the time coordinate, that is, if $\mathbf{L}$ does not depend on t explicitly, then $\mathbf{L}_t = 0$, and we identify the homogeneous form of (2.116) with the energy conservation equation. This result is also a consequence of Noether's theorem. Thus, $\sigma\mathbf{L}_\sigma$ is the energy density E and $-\sigma\mathbf{L}_{k_i}$ is the energy flux. For small amplitude waves, $\mathbf{L} \equiv 0$, yielding

$$\sigma\mathbf{L}_\sigma = E, \qquad (2.117)$$

where E is the average energy density of the waves. Hence, the wave-action given by

$$\mathbf{L}_\sigma = \frac{E}{\sigma} \qquad (2.118)$$

is conserved for small amplitude waves propagating in a nonuniform, time-dependent, but stationary $\mathbf{U} = 0$; that is, media at rest, through the wave-action conservation equation (2.115)

$$\frac{\partial}{\partial t}\left(\frac{E}{\sigma}\right) + \frac{\partial}{\partial x_i}\left(C_{gi}\frac{E}{\sigma}\right) = 0. \qquad (2.119)$$

The special case is that if σ is invariant, or as shown, the medium is time independent and stationary, then

$$\frac{\partial E}{\partial t} + \frac{\partial}{\partial x_i}(C_{gi}E) = 0. \tag{2.120}$$

Therefore, conservation of wave-action is more fundamental than conservation of energy. The conservation of wave-action holds in a frame of reference at rest with respect to the fluid, whereas the conservation of energy holds only in time-independent, stationary media.

The result of wave-action conservation can be generalized to the frame of reference in which the fluid is local at rest and may be moving with a slowly varying velocity $\mathbf{U}$ with respect to an inertial coordinate system, that is, it is in moving media. In this case, the wave-action can be shown to be in the following form:

$$L_\sigma = \frac{E}{\sigma_0}, \tag{2.121}$$

and the wave-action conservation becomes

$$\frac{\partial}{\partial t}\left(\frac{E}{\sigma_0}\right) + \nabla \cdot \left(\mathbf{C_g}\frac{E}{\sigma_0}\right) = 0 \tag{2.122}$$

(see Bretherton and Garrett, 1968; Garrett, 1968), where again the subscripts 0, as mentioned earlier, refer to quantities observed within the moving system. For example, the frequency is the intrinsic frequency, that is, the frequency measured in a frame of reference moving with the local basic flow $\mathbf{U}$.

2.5.6 THE ENERGY EXCHANGE WITH THE BASIC FLOW

If there are no interactions between the wave packet and the basic state of medium, that is, basic flow, then the energy of a wave packet would be conserved and the energy equation would be eq. (2.120). However, in general, the interaction does exist. The equation (2.122) may be rewritten as

$$\frac{D_g E}{Dt} + E\nabla \cdot \mathbf{C}_g - \frac{E}{\sigma_0}\frac{D_g\sigma_0}{Dt} = 0, \tag{2.123}$$

where

$$\frac{D_g}{Dt} \equiv \frac{\partial}{\partial t} + \mathbf{C}_g \cdot \nabla \tag{2.124}$$

is derivative along the characteristic curve, the same as (2.92). Hence, the $-(E/\sigma_0)(D_g\sigma_0/Dt)$ in the same way represents the interaction between the wave packet and the mean flow. But $D_g\sigma_0/Dt$ is given by

$$\frac{D_g\sigma_0}{Dt} = k_j C_{gi0}\frac{\partial U_j}{\partial x_i} + \frac{\partial \sigma_0}{\partial B_j}\frac{DB_j}{Dt}, \tag{2.125}$$

from (2.98) by use of (2.95), where

$$\frac{D}{Dt} \equiv \frac{\partial}{\partial t} + \mathbf{U} \cdot \nabla \tag{2.126}$$

is derivative along the basic flow. Therefore, (2.123) becomes

$$\frac{D_g E}{Dt} + E \nabla \cdot \mathbf{C}_g + \frac{E k_j}{\sigma_0} C_{gi0} \frac{\partial U_j}{\partial x_i} - \frac{E}{\sigma_0} \frac{\partial \sigma_0}{\partial B_j} \frac{DB_j}{Dt} = 0. \tag{2.127}$$

For most types of wave motion, it is usually possible to relate the variations of the basic state, as expressed by derivative of B_j, to the current shear by an equation of the form (Garrett, 1968)

$$\frac{1}{B_j} \frac{DB_j}{Dt} + \Lambda_{ij} \frac{\partial U_j}{\partial x_i} = 0, \tag{2.128}$$

where Λ_{ij} is a tensor independent of $\mathbf{U}$ and its derivatives, and where, in the first term, the conventional summation should not be applied. Thus, the energy equation in a moving flow (2.127), may then be recast as

$$\frac{\partial E}{\partial t} + \nabla \cdot (\mathbf{C}_g E) + T_{ij} \frac{\partial U_j}{\partial x_i} = 0, \tag{2.129}$$

and the rate of loss of wave energy within a volume, each point of which moves with the group velocity, may be equated to the rate of working, against the rate of strain of the basic flow, of an *interaction stress tensor* T_{ij}, given by

$$T_{ij} = \frac{E}{\sigma_0} \left(B_k \frac{\partial \sigma_0}{\partial B_k} \Lambda_{ij} + C_{gi0} k_j \right). \tag{2.130}$$

For particular types of wave propagation, the components of T_{ij} occurring in the energy equation for waves in a basic flow and satisfying the requirements of slow variation may be identified with the corresponding components of the *radiation stress tensor* for the waves, which is defined as the second order mean of the flux.

2.6 The General Wave-Action Equation in the GLM Description: Andrews and McIntyre's Theory

The *Generalized Lagrangian-mean* (GLM) is an important concept in the study of wave theory, especially in the theory of wave-mean interaction. This theory was introduced and developed by Andrews and McIntyre (1978b,c) a decade ago. The theory is introduced here not only because it is the most fundamental wave-action equation, which certainly provides a much better understanding of wave-action discussed in the previous section, but also because the theory has already had a very important impact on studies of wave theory and wave-mean interaction, especially those in geophysical fluid dynamics.

An exact and very general Lagrangian-mean description of the back effect of the wave or the wave packet upon the mean flow had been studied using the theory. The basic formalism can be applied to any problems in which governing equations are given in the usual Euler–Lagrange form, as given in (2.100) and irrespective of whether spatial, temporal, ensemble, or *two-timing* averages are appropriate. The equations embody generalizations of the *Eliassen–Palm theorem* (Eliassen and Palm, 1961) and the *Charney–Drazin theorem* (Charney and Drazin, 1961) in applications in geophysical fluid dynamics. However, we will not discuss those results here, and the reader who wishes to know these aspects can consult the papers by Andrews and McIntyre (1976a,b; 1978a,b,c) or a rewiew by McIntyre (1977). Applications of GLM theory to stratified shear flows were reviewed by Grimshaw (1984).

Here, we begin with the generalized Lagrangian-mean description (Section 2.6.1). Then, a general wave-action equation in GLM description is studied. The basic theorems of mean-flow is given in Appendix A.

2.6.1 THE GENERALIZED LAGRANGIAN-MEAN DESCRIPTION

Define the *Eulerian-mean operator* $\overline{(\ )}$, a general averaging operator taking any field $\phi(\mathbf{x}, t)$, into its corresponding Eulerian-mean field $\bar{\phi}(\mathbf{x}, t)$. Therefore, the Eulerian-mean operator $\overline{(\ )}$ is the averaging in the general sense.

We define $\overline{(\ )}^L$ as the *generalized Lagrangian-mean operator,* which corresponds to any given Eulerian-mean operator $\overline{(\ )}$. The field $\phi(\mathbf{x}, t)$ in the generalized Lagrangian-mean operator is written as

$$\overline{\phi(\mathbf{x}, t)}^L = \overline{\phi[\mathbf{x} + \xi(\mathbf{x}, t)]}, \qquad (2.131)$$

where $\xi(\mathbf{x}, t)$ is the disturbance-associated particle displacement field. Hence, $\overline{(\ )}^L$ is defined as the average taken with respect to the displacement position $\mathbf{x} + \xi$, and considered as a function of space and of time.

We can recall two properties of the mapping,

$$\mathbf{x} \rightarrow \mathbf{x} + \xi(\mathbf{x}, t), \qquad (2.132)$$

for a general function $\xi(\mathbf{x}, t)$, regarding it simply as a general transformation in the spirit of Eckart (1963) and Soward (1972). Introducing the notation

$$\phi^\xi(\mathbf{x}, t) = \phi[\mathbf{x} + \xi(\mathbf{x}, t), t], \qquad (2.133)$$

we have identities

$$(\phi^\xi)_{,t} = (\phi_{,t})^\xi + (\phi^\xi_{,j})\Xi_{j,t} \tag{2.134a}$$

and

$$(\phi^\xi)_{,i} = (\phi_{,j})^\xi \Xi_{j,i}, \tag{2.134b}$$

where

$$\Xi(\mathbf{x}, t) = \mathbf{x} + \xi(\mathbf{x}, t), \tag{2.135}$$

$$( )_{,t} = \frac{\partial( )}{\partial t}, \quad \text{and} \quad ( )_{,i} = \frac{\partial( )}{\partial x_i}.$$

We assume that the mapping (2.132) is invertible. Then, for any given $\mathbf{u}(\mathbf{x}, t)$, there is evidently a unique *related velocity* $\mathbf{v}(\mathbf{x}, t)$ such that when the point $\mathbf{x}$ moves with velocity $\mathbf{v}$ the $\mathbf{x} + \xi$ moves with actual fluid velocity $\mathbf{u}^\xi$, that is,

$$\left(\frac{\partial}{\partial t} + \mathbf{v} \cdot \nabla\right)\Xi = \mathbf{u}^\xi. \tag{2.136}$$

An immediate consequence is

$$\left(\frac{\partial}{\partial t} + \mathbf{v} \cdot \nabla\right)(\phi^\xi) = \left(\frac{D\phi}{Dt}\right)^\xi, \tag{2.137}$$

for any field ϕ, where

$$\frac{D}{Dt} = \frac{\partial}{\partial t} + \mathbf{u} \cdot \nabla. \tag{2.138}$$

The generalized Lagrangian-mean description requires

$$\overline{\xi(\mathbf{x}, t)} = 0 \tag{2.139}$$

and

$$\overline{\mathbf{v}(\mathbf{x}, t)} = \mathbf{v}(\mathbf{x}, t). \tag{2.140}$$

Thus, ξ is a true disturbance quantity and $\mathbf{v}$ is a mean quantity in the GLM description.

By virtue of the above definition, and the fact that $\mathbf{v} = \bar{\mathbf{u}}^L$, we have the *Lagrangian-mean material derivative*

$$\bar{D}^L = \frac{\partial}{\partial t} + \bar{\mathbf{u}}^L \cdot \nabla. \tag{2.141}$$

Then eq. (2.136) may be written as

$$\bar{D}^L \Xi = \mathbf{u}^\xi, \tag{2.142a}$$

or equivalently,

$$\bar{D}^L \xi = \mathbf{u}^\ell, \tag{2.142b}$$

where

$$\mathbf{u}^\ell(\mathbf{x}, t) = \mathbf{u}^\xi - \bar{\mathbf{u}}^L. \tag{2.142c}$$

The quantity $\mathbf{u}^\ell$ is called the *Lagrangian disturbance velocity*.

A further postulate is introduced in order to formally define ξ and $\bar{\mathbf{u}}^L$ for the general $\overline{(\ )}$. That is, each Lagrangian-mean trajectory [solution $d\mathbf{x}/dt = \bar{\mathbf{u}}^L(\mathbf{x}, t)$] passes through at least one point $(\mathbf{x}_0, t_0)$ in the neighborhood of which there is no disturbance. That is to say, there is a point $(\mathbf{x}_0, t_0)$ at which

$$\xi\big|_{\mathbf{x}=\mathbf{x}_0, t=t_0} = 0. \tag{2.143}$$

From the above, it can easily be seen that in the generalized Lagrangian-mean description, the averaging results are very simple. Summarizing the above results, we have

$$\left(\frac{D\phi}{Dt}\right)^\xi = \bar{D}^L(\phi^\xi), \tag{2.144}$$

from (2.137). Because the velocity field involved in $\bar{D}^L$ is a mean field two important corollaries can be derived from (2.144):

$$\overline{\left(\frac{D\phi}{Dt}\right)}^L = \bar{D}^L \bar{\phi}^L \tag{2.145}$$

and

$$\left(\frac{D\phi}{Dt}\right)^\ell = \bar{D}^L \phi^\ell, \tag{2.146}$$

where

$$\phi^\ell = \phi^\xi - \bar{\phi}^L, \tag{2.147a}$$
$$\bar{\phi}^\ell = 0, \tag{2.147b}$$
$$\bar{\phi}^L \equiv \bar{\phi}^\xi. \tag{2.147c}$$

Compared with the results in the Eulerian description, that is,

$$\frac{\overline{D\phi}}{Dt} = \bar{D}\bar{\phi} + \overline{\mathbf{u}' \cdot \nabla \phi'} \tag{2.148}$$

and

$$\left(\frac{D\phi}{Dt}\right)' = \bar{D}\phi' + \mathbf{u}' \cdot \nabla\bar{\phi} + \mathbf{u} \cdot \nabla\phi' - \overline{\mathbf{u}' - \nabla\phi'}, \tag{2.149}$$

where

$$\bar{D} = \frac{\partial}{\partial t} + \bar{\mathbf{u}} \cdot \nabla, \tag{2.150}$$

that is, a derivative along the Eulerian-mean, and

$$\phi' = \phi - \bar{\phi} \tag{2.151}$$

(the Eulerian disturbance field). Therefore, the results show that the GLM description is much simpler than the Eulerian-mean description.

The two important basic theorems with several corollaries for the mean-flow evolution can be derived, and are provided, in Appendix A. The proof and implications of the theorems are found in the original paper.

2.6.2   THE GENERAL WAVE-ACTION EQUATION

Before trying to apply the GLM description to geophysical fluids, we can rewrite eq. (1.5) in the following form:

$$\frac{Du_i}{Dt} + 2(\Omega \times \mathbf{u})_i + \Phi_{,i} + \rho^{-1}p_{,i} + F_i = 0 \qquad (2.152)$$

for the ith component, where Φ is a potential for the sum of the gravitational and centrifugal forces, and the other terms are defined as before.

Now, let us derive the general wave-action equation from eq. (2.152), by multiplying (2.152) by $\xi_{i,a}$.

First, we note

$$\overline{\phi_{,\alpha}} = 0, \qquad (2.153)$$

so that

$$\overline{\phi_{,\alpha}\psi} = -\overline{\phi\psi_{,\alpha}} \qquad (2.154)$$

for any ϕ and ψ. Also for any mean field, $\bar{\phi}$, we have

$$[(\bar{\phi})^\xi]_{,\alpha} = (\overline{\phi_{,i}})^\xi(x_i + \xi_i)_{,\alpha} = (\overline{\phi_{,\alpha}})^\xi \xi_{i,\alpha}, \qquad (2.155)$$

so that

$$\overline{(\overline{\phi_{,i}})^\xi \xi_{i,\alpha}} = 0. \qquad (2.156)$$

The relations (2.147) and (2.144), applied to $\mathbf{u}$, read

$$\mathbf{u}^\xi = \bar{\mathbf{u}}^L + \mathbf{u}^\ell, \qquad (2.157a)$$

$$\overline{\mathbf{u}^\ell} = 0, \qquad (2.157b)$$

and

$$\left(\frac{D\mathbf{u}}{Dt}\right)^\xi = \bar{D}^L(\mathbf{u}^\xi). \qquad (2.158)$$

Multiplying the first term of (2.152) by $\xi_{i,\alpha}$ and averaging and using (2.147a,b) and (2.158), we obtain

$$\overline{\xi_{i,\alpha}\bar{D}^L u_i^\xi} = \overline{\xi_{i,\alpha}\bar{D}^L u_i^\ell}$$
$$= \bar{D}^L\overline{(\xi_{i,\alpha}u_i^\ell)} - \overline{u_i^\ell \bar{D}^L \xi_{i,\alpha}}$$
$$= \bar{D}^L\overline{(\xi_{i,\alpha}u_i^\ell)}, \qquad (2.159)$$

where (2.153) and (2.142) have been used. The Coriolis term gives

$$2\Omega_j\varepsilon_{ijk}\overline{\xi_{i,\alpha}u_k^{\ell}} = 2\Omega_j\varepsilon_{ijk}\overline{\xi_{i,\alpha}\bar{D}^L\xi_k}$$
$$= -2\Omega_j\varepsilon_{ijk}\overline{\xi_{k,\alpha}\bar{D}^L\xi_i}, \tag{2.160a}$$

where ε_{ijk} is the alternating tensor and $\varepsilon_{ijk} = -\varepsilon_{kji}$. This term can be further written as

$$2\Omega_j\varepsilon_{ijk}\overline{\xi_k\bar{D}^L\xi_{i,\alpha}}, \tag{2.160b}$$

by the use of (2.154). The term then becomes

$$\Omega_j\varepsilon_{ijk}\bar{D}^L\overline{(\xi_{i,\alpha}\xi_k)}. \tag{2.160c}$$

The term $\Phi_{,i}$ gives zero by the relation (2.156).

As far as the pressure term is concerned, we have to note first that

$$(\phi^{\xi})_{,j} = (\phi_{,k})^{\xi}(\delta_{kj} + \xi_{k,j}), \tag{2.161}$$

where δ_{kj} is the delta function.

Define the Jacobian J of the mapping (2.132) as

$$J \equiv \det(\delta_{ij} + \xi_{i,j}), \tag{2.162a}$$

to satisfy

$$\tilde{\rho} \equiv \rho^{\xi}J = \bar{\bar{\rho}}, \tag{2.162b}$$

where $\tilde{\rho}$ is the mean quantity of density and also satisfies the mean-flow mass conservation

$$\bar{D}^L\tilde{\rho} + \tilde{\rho}\nabla \cdot \bar{\mathbf{u}}^L = 0. \tag{2.163}$$

Furthermore, define K_{ij} as the (i,j)th cofactor of the Jacobian J, to satisfy

$$(\delta_{jk} + \xi_{j,k})K_{ji} = J\delta_{ik}, \tag{2.164a}$$
$$= (\delta_{kj} + \xi_{k,j})K_{ij}, \tag{2.164b}$$

which yields

$$K_{ij,j} = 0 \tag{2.165}$$

and

$$K_{ij} = (1 + \xi_{m,m})\delta_{ij} - \xi_{j,i} + k_{ij}, \tag{2.166}$$

where k_{ij} is the (i,j)th cofactor of ξ_{ij}.

On multiplying (2.161) by K_{ij}/J and using (2.164b), we obtain the inverse relation

$$(\phi_{,i})^{\xi} = \frac{(\phi^{\xi})_{,j}K_{ij}}{J}. \tag{2.167}$$

Hence, by using (2.162), we obtain the pressure term

$$\overline{\xi_{i,\alpha}(\rho^{-1}p_{,i})^{\xi}} = \tilde{\rho}^{-1}\overline{(p^{\xi})_{,j}\xi_{i,\alpha}K_{ij}}. \tag{2.168}$$

Further, it reads

$$\tilde{\rho}^{-1}\overline{(p^\xi\xi_{i,\alpha}K_{ij})_{,j}} - \tilde{\rho}^{-1}\overline{p^\xi J_{,\alpha}}, \qquad (2.169)$$

in virtue of (2.165), since

$$J_{,\alpha} = K_{ij}\xi_{i,j\alpha}. \qquad (2.170)$$

The equation (2.169) can also be written as

$$\tilde{\rho}^{-1}\overline{(p^\xi\xi_{i,\alpha}K_{ij})_{,j}} + \overline{(p^\xi)_{,\alpha}/\rho^\xi}, \qquad (2.171)$$

by the use of (2.154) and (2.162).

Therefore, summarizing the above results, multiplying (2.152) by $\xi_{,\alpha}$, and averaging, we obtain

$$\bar{D}^L A + \tilde{\rho}^{-1}\nabla \cdot \mathbf{B} = \tilde{F}, \qquad (2.172)$$

where

$$A = \overline{\xi_{,\alpha} \cdot (\mathbf{u}^L + \mathbf{\Omega} \times \boldsymbol{\xi})} \qquad (2.173)$$

is the *wave-action* per unit bass and $\mathbf{B}$ is the nonadvective flux of wave-action, defined by

$$B_j \equiv \overline{p^\xi\xi_{i,\alpha}K_{ij}}. \qquad (2.174)$$

The right-hand side of (2.172) is given by

$$\tilde{F} \equiv -\overline{\xi_{i,\alpha}F_i^\ell} + \overline{(p^\ell)_{,\alpha}q}, \qquad (2.175)$$

a wave property representing the rate of generation or dissipation of wave-action associated with departures from conservative motion, since it is zero for the conservation motion, q, in particular, being the departure from adiabatic motion, since it is zero for that motion. It is defined by

$$q = -\frac{1}{\rho(S^\xi, p^\xi)} + \frac{1}{\rho(\bar{S}^L, p^\xi)}, \qquad (2.176)$$

where

$$\rho = \rho(S, p) \qquad (2.177)$$

is the equation of state for the fluid and S is the entropy per unit mass.

The wave-action equation (2.172) also may be written as

$$\frac{\partial(\tilde{\rho}A)}{\partial t} + \nabla \cdot \mathbf{B}^{tot} = \tilde{\rho}\tilde{F}, \qquad (2.178)$$

by the use of the mean-flow mass conservation equation (2.163), where $\mathbf{B}^{tot}$ is the total flux of wave-action, defined as

$$\mathbf{B}^{tot} = \bar{\mathbf{u}}^L\tilde{\rho}A + \mathbf{B}, \qquad (2.179)$$

the first term of which represents the advection of wave-action by the mean flow $\bar{\mathbf{u}}^L$.

Note: Please be aware that the derivation of the wave-action equation (4.14) in the original paper (Andrews and McIntyre (1978c)) is incorrect, as pointed out by Andrews and McIntyre (1979).

2.7 Closure

In writing this book, the author found it difficult to organize the material in this chapter. Fortunately, there have been many books on wave theory. It is recommended that readers refer to these books. They mainly consider propagation properties and stabilities of waves. The book by Whitham (1974) details the development of both linear and nonlinear waves, in general. The book by Lighthill (1978) studies waves in fluids. A recent book by Craik (1985) discusses wave interactions in the fluid flows and the general problems of wave and mean flows, including modulated wave packets and the GLM formulation.

Appendix A. Basic Theorems of Mean-Flow in the GLM Description

Let $T(S,p)$ and $\rho(S,p)$ represent the temperature and density in thermodynamics, which depend upon the entropy S per unit mass and the pressure p. Similarly, $H(S,p)$ represents the enthalpy per unit mass. The nonlinear forcing of the mean flow by the waves is expressed in terms of a vector wave property, $\mathbf{P}$, whose component is

$$P_i(x,t) = -\overline{\xi_{j,i}[u_j^\ell + (\mathbf{\Omega} \times \mathbf{\xi})_j]}. \tag{2.180}$$

It can be proven:

Theorem I

$$\bar{D}^L(\bar{u}_i^L - P_i) + (\bar{u}_k^L)_{,i}(\bar{u}_k^L - P_k) + 2(\mathbf{\Omega} \times \bar{\mathbf{u}}^L)_i + \Pi_{,i}$$
$$- (\bar{S}^L)_{,i} T(\bar{S}^L; \rho^\xi) + \bar{F}_i^L = -\overline{\xi_{j,i} F_j^\ell} + \overline{(p^\xi)_{,i} q}, \tag{2.181}$$

where

$$\Pi \equiv \overline{H(\bar{S}^L, p^\xi)} + \bar{\Phi}^L - \overline{u_j^\xi\left[\frac{1}{2}u_j^\xi + (\mathbf{\Omega} \times \mathbf{\xi})_j\right]} \tag{2.182}$$

and q is given in (2.171), which is zero for the adiabatic motion.

Corollary I. *If all mean quantities are independent of the Cartesian coordinate x_i, then*

$$\bar{D}^L(\bar{\mathbf{u}}_i^L - P_i) + 2(\mathbf{\Omega} \times \bar{\mathbf{u}}^L)_i + \bar{F}_i^L = -\overline{\xi_{j,i} F_j^\ell} + \overline{(p^\ell)_{,i} q}. \tag{2.183}$$

Corollary II. *If all mean quantities are axisymmetric about an axis through the origin of coordinates and parallel to $\mathbf{\Omega}$, then*

$$\bar{D}^L[\hat{z} \cdot \mathbf{x} \times (\bar{\mathbf{u}}^L - \mathbf{P})] + 2\Omega \mathbf{x}^\perp \cdot \bar{\mathbf{u}}^L + \hat{z} \cdot \mathbf{x} \times \bar{\mathbf{F}}^\mathbf{L}$$
$$= \varepsilon_{ijk}\hat{z}_j x_k\left[-\overline{\xi_{m,i} F_m^\ell} + \overline{(p^\ell)_{,i} q}\right], \tag{2.184}$$

which can be obtained by taking $\hat{z} \cdot \mathbf{x} \times$ (2.181), where $\hat{z}$ is a unit vector parallel to $\mathbf{\Omega}$, and

$$\mathbf{x}^{\perp} = \mathbf{x} - (\hat{z} \cdot \mathbf{x})\hat{z}. \tag{2.185}$$

Corollary III. *Let Γ be a closed circuit moving with velocity $\bar{\mathbf{u}}^L$. Then,*

$$\frac{d}{dt} \oint_{\Gamma} (\bar{\mathbf{u}}^L - \mathbf{P} + \mathbf{\Omega} \times \mathbf{x}) \cdot d\mathbf{s} - \oint_{\Gamma} \overline{T(\bar{S}^L, p^{\xi})} d\bar{S}^L$$

$$= - \oint_{\Gamma} \left[F_i^L + \overline{\xi_{j,i} F_j^{\ell}} - \overline{(p^{\xi})_{,i} q} \right] ds_i. \tag{2.186}$$

Corollary IV. *If $\mathbf{\Omega} = 0$ and the motion is irrotational, homentropic ($S = constant$), and conservative ($\mathbf{F} = 0$, $q = 0$), then $\bar{\mathbf{u}}^L - \mathbf{P}$ is irrotational.*

Theorem II

$$\bar{D}^L \left(\frac{1}{2} \overline{|\mathbf{u}^{\xi}|^2} + \bar{H}^L + \bar{\Phi}^L - e \right) - (\bar{u}_k^L)_{,t}(\bar{u}_k^L - P_k) - \Pi_{,t}$$

$$+ (\bar{S}^L)_{,t}\overline{T(\bar{S}^L, p^{\xi})} + \bar{\mathbf{u}}^L \cdot \overline{\mathbf{F}^L} + \overline{(TQ)}^L$$

$$= -\bar{u}_k^L \overline{\xi_{j,k} F_j^{\ell}} - \overline{(p^{\xi})_{,t} q}, \tag{2.187}$$

where the Lagrangian-mean enthalpy

$$\bar{H}^L = \overline{H(S, p)}^L = \overline{H(S^{\xi}, p^{\xi})}, \tag{2.188}$$

$$e = \overline{\xi_{j,t}[u_j^L + (\mathbf{\Omega} \times \xi)_j]}, \tag{2.189}$$

and Q are the sources and sinks of the entropy S.

References

Andrews, D.G., and McIntyre, M.E. (1976a). Planetary waves in horizontal and vertical shear: The generalized Eliassen–Palm relation and the mean zonal acceleration. *J. Atmos. Sci.* **33**, 2031–2048.

Andrews, D.G., and McIntyre, M.E. (1976b). Planetary waves in horizontal and vertical shear: Asymptotic theory for equatorial waves in weak shear. *J. Atmos. Sci.* **33**, 2049–2053.

Andrews, D.G., and McIntyre, M.E. (1978a). Generalized Eliassen–Palm and Charney–Drazin theorems for waves on axisymmetric flows in compressible atmospheres. *J. Atmos. Sci.* **35**, 175–185.

Andrews, D.G., and McIntyre, M.E. (1978b). An exact theory of nonlinear waves on a Lagrangian-mean flow. *J. Fluid Mech.* **89**, 609–646.

Andrews, D.G., and McIntyre, M.E. (1978c). On wave-action and its relatives. *J. Fluid Mech.* **89**, 647–664.

Andrews, D.G., and McIntyre, M.E. (1979). On wave-action and its relatives. *J. Fluid Mech.* **95**, 795.

Arnold, V.I. (1978). *Mathematical Methods of Classical Mechanics.* Springer-Verlag, New York (Russian original, Moscow, 1974).

Bretherton, F.P., and Garrett, C.J.R. (1968). Wavetrains in inhomogeneous moving media. *Proc. Roy. Soc. London* **A302**, 529–554.

Charney, J.G., and Drazin, P.G. (1961). Propagation of planetary-scale disturbances from the lower into upper atmosphere. *J. Geophys. Res.* **66**, 83–109.

Courant, R., and Hilbert, D. (1962). *Methods of Mathematical Physics*, Vol. II. John Wiley, New York.

Craik, A.D.D. (1985). *Wave Interactions and Fluid Flows.* Cambridge University Press, London.

Dennery, P., and Krizywicki (1967). *Methods of Mathematical Physics.* Harper and Row, New York.

Eckart, C. (1963). Some transformations of the hydrodynamical equations. *Phys. Fluids* **6**, 1037–1041.

Eliassen, A., and Palm, E. (1961). On the transfer of energy in stationary mountain waves. *Geophys. Publ.* **22**(3), 1–23.

Garrett, C.J.R. (1968). On the interaction between internal gravity waves and a shear flow. *J. Fluid Mech.* **34**, 711–720.

Gelfand, I.M., and Fomin, S.V. (1963). *Calculus of Variations.* Prentice-Hall, Englewood Cliffs, N.J.

Goldstein, H. (1980). *Classical Mechanics.* 2nd ed. Addison-Wesley, Reading, Mass.

Grimshaw, R.H.J. (1984). Wave action and wave-mean flow interaction, with application to stratified shear flows. *Ann. Rev. Fluid Mech.* **16**, 11–44.

Hayes, W.D. (1970). Conservation of action and modal wave action. *Proc. Roy. Soc. London* **A320**, 187–208.

Jeffreys, H., and Jeffreys, B.S. (1956). *Methods of Mathematical Physics.* Cambridge University Press, London.

Keller, J.B. (1958). Surface waves on non-uniform depth. *J. Fluid Mech.* **4**, 607–614.

Lighthill, M.J. (1965). Group velocity. *J. Ins. Math. Appl.* **1**, 1–28.

Lighthill, M.J. (1978). *Waves in Fluids.* Cambridge University Press, London.

McIntyre, M.E. (1977). Wave transport in stratified, rotating fluids. *Lecture Notes in Physics* No. 71. Springer-Verlag, New York.

Soward, A.M. (1972). A kinematic theory of large magnetic Reynolds number dynamos. *Phil. Trans. Roy. Soc. London* **A272**, 431–462.

Whitham, G.B. (1965a). Nonlinear dispersive waves. *Proc. Roy. Soc. London* **A283**, 238–261.

Whitham, G.B. (1965b). A general approach to linear and nonlinear dispersive waves using a Lagrangian. *J. Fluid Mech.* **22**, 273–283.

Whitham, G.B. (1974). *Linear and Nonlinear Waves*. John Wiley, New York.

3

Evolution of the Wave Packet in Barotropic Flows

3.1 Introduction

We now are ready to apply the theory developed in Chapter 2 to geophysical fluid dynamics. In this chapter and the next three chapters, we consider the evolution of the wave packet, specifically the Rossby wave packet, only in barotropic basic flows in order to clarify ideas, demonstrate the effectiveness of the theory, and show some new and fundamental results as well. The case in which the stratification and baroclinicity in the basic flows are both taken into account is postponed to Chapter 7.

In this chapter, we first give the inviscid shallow-water model and derive the potential vorticity equation in Section 3.2. Then, by use of the theory developed in Chapter 2 with the theorems, we form a set of equations governing the evolution of the Rossby wave packet, that is, eqs. (2.93b,c) and (2.90), which, in this case, are similar to eqs. (2.76), (2.77a,b), and (2.78a,b). After that, the integral properties of the wave packet in the zonal basic flow is discussed. The three barotropic instability theorems in the wave packet are presented. They are shown to be analogous to the previous finding in barotropic instability associated with the wave-mean interaction (e.g., Pedlosky, 1987) and the other two to the Rayleigh theorem and the Fjørtoft theorem (Drazin and Reid, 1981). The roles of the earth's rotation, basic flow, and topography in the structural change of the wave packet in geophysical fluids is systematically discussed in Section 3.5. The physical explanation for the mechanism of the structural change is given in Section 3.6, with two examples. And the final section is a summary of the results.

3.2 The Inviscid, Shallow-Water Model and Potential Vorticity Equation

The equations governing the inviscid, incompressible, shallow-water model, as shown in Figure 3.1, can be written as follows:

$$\frac{\partial u}{\partial t} + u \frac{\partial u}{\partial x} + v \frac{\partial u}{\partial y} - fv = -g \frac{\partial h}{\partial x}, \qquad (3.1)$$

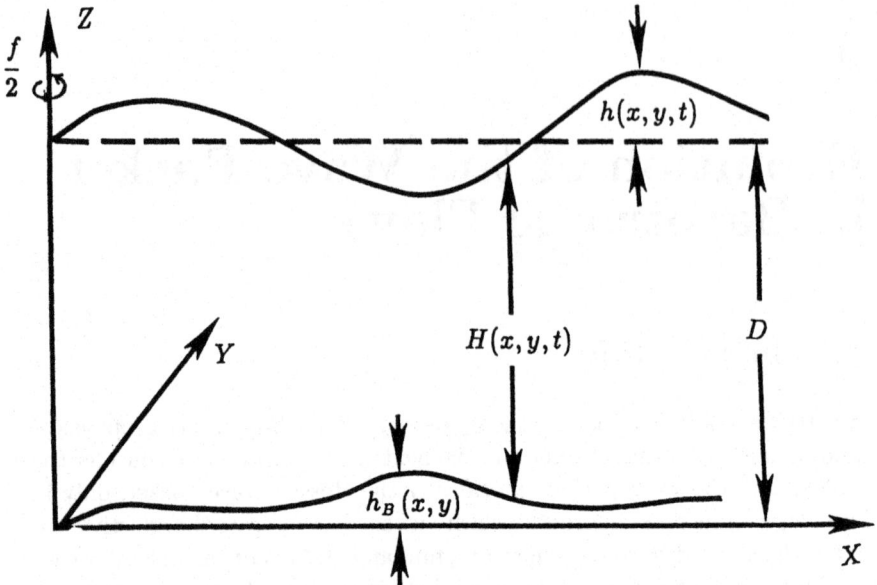

FIGURE 3.1. Shallow-water model with topography.

$$\frac{\partial v}{\partial t} + u\frac{\partial v}{\partial x} + v\frac{\partial v}{\partial y} + fu = -g\frac{\partial h}{\partial y}, \qquad (3.2)$$

and

$$\frac{dH}{dt} + H\left(\frac{\partial u}{\partial x} + \frac{\partial v}{\partial y}\right) = 0, \qquad (3.3)$$

from (1.5) and (1.3), that is, the momentum equation and the continuity equation, where

$$H(x, y, t) = D + h(x, y, t) - h_B(x, y), \qquad (3.4)$$

$h_B(x, y)$ is the height of the topography, which is the only function of space, and $h(x, y, t)$ is the relative height of the free surface of the fluid.

The scaling for the variables is as follows:

$$(x, y) = L(x', y'), \qquad t = \tau t', \qquad (3.5a)$$
$$(u, v) = U(u', v'), \qquad gh = N_0\psi', \qquad (3.5b)$$

and

$$H = h_0(x, y) + \psi'(x, y, t)$$
$$= D + h(x, y, t) - h_B(x, y), \qquad (3.6)$$

where D is the average height of the model.

On the δ-surface approximation (1.24), we can write the Coriolis parameter in the following form:

$$f \simeq f_0\left(1 + \frac{\beta}{f_0}Ly' - \frac{L^2}{a^2}y'^2\right)$$
$$= f_0\left(1 + \bar{\beta}y' - \frac{\delta}{2}y'^2\right), \tag{3.7}$$

where

$$\bar{\beta} = \frac{\beta}{f_0}L, \qquad \delta = \frac{L^2}{a^2}. \tag{3.8}$$

Substituting (3.5), (3.6), and (3.8) into (3.1) to (3.3) after nondimensionization we obtain

$$\varepsilon_T\frac{\partial u}{\partial t} + \varepsilon\left(u\frac{\partial u}{\partial x} + v\frac{\partial u}{\partial y}\right) - \left(1 + \bar{\beta}y - \frac{\delta}{2}y^2\right)v = -\frac{\partial\psi}{\partial x}, \tag{3.9}$$

$$\varepsilon_T\frac{\partial v}{\partial t} + \varepsilon\left(u\frac{\partial v}{\partial x} + v\frac{\partial v}{\partial y}\right) + \left(1 + \bar{\beta}y - \frac{\delta}{2}y^2\right)u = -\frac{\partial\psi}{\partial y}, \tag{3.10}$$

and

$$\varepsilon_T F\frac{\partial\psi}{\partial t} + \varepsilon F\left(u\frac{\partial\psi}{\partial x} + v\frac{\partial\psi}{\partial y}\right) - u\frac{\partial}{\partial x}\left(\frac{h_B}{D}\right) - v\frac{\partial}{\partial y}\left(\frac{h_B}{D}\right)$$
$$+ \left(1 + \varepsilon F\psi - \frac{h_B}{D}\right)\left(\frac{\partial u}{\partial x} + \frac{\partial v}{\partial y}\right) = 0, \tag{3.11}$$

where

$$\varepsilon_T = \frac{U}{f_0\tau}, \tag{3.12}$$

$$\varepsilon = \frac{U}{f_0 L}, \tag{3.13}$$

$$F = \frac{f_0^2 L^2}{gD} = \left(\frac{L}{R}\right)^2, \tag{3.14}$$

and

$$R = \sqrt{\frac{gD}{f_0}}, \tag{3.15}$$

and ε_T, ε, and F are the dimensionless parameters. ε is called the *Rossby number*, and F is the *Froude number*. R is the *radius of the Rossby deformation*. In eqs. (3.9) to (3.11), the primes have been dropped.

For the synoptic scale system in the atmosphere,

$$L \sim R \sim 10^3\,\text{km}, \tag{3.16}$$

and

$$\varepsilon \sim 10^{-1}, \tag{3.17}$$

and we let the characteristic time equal the advection time. Then,

$$\frac{\varepsilon T}{\varepsilon} \sim 1. \tag{3.18}$$

For the terms in the Coriolis parameter, we have

$$\frac{\delta}{\bar{\beta}} = \frac{L}{a} \tan \phi_0, \tag{3.19}$$

when

$$\phi_{0c} = \arctan \frac{a}{L} \tag{3.20}$$

and

$$\delta \sim \bar{\beta}. \tag{3.21}$$

If $L \sim 3.5 \times 10^3$ km and $a \simeq 6.7 \times 10^3$ km, then $\phi_{0c} \simeq 62°$. Therefore, the effect of δ is greater than β, when $\phi > \phi_{0c}$, smaller than β when $\phi < \phi_{0c}$, and equal to β when $\phi = \phi_{0c}$. However, for the ultralong scale systems, the two factors have the same order of magnitude and are of equal importance. Suppose that the β and δ both have the order of magnitude of ε, that is,

$$\bar{\beta} \sim \varepsilon \beta_0, \qquad \delta \sim \varepsilon \bar{\delta}. \tag{3.22}$$

In addition, for the low topography, we assume

$$\frac{h_B}{D} = \varepsilon \eta_B(x, y). \tag{3.23}$$

Perturbing variables u, v, and ψ in a series of a small parameter, ε, we have

$$u = u_0 + \varepsilon u_1 + \varepsilon^2 u_2 + \cdots,$$
$$v = v_0 + \varepsilon v_1 + \varepsilon^2 u_2 + \cdots, \tag{3.24}$$
$$\psi = \psi_0 + \varepsilon \psi_1 + \varepsilon^2 \psi_2 + \cdots.$$

Substituting (3.22) to (3.24) into eqs. (3.9) to (3.11), we obtain the zeroth order equations

$$-v_0 = -\frac{\partial \psi_0}{\partial x}, \tag{3.25}$$

$$u_0 = -\frac{\partial \psi_0}{\partial y}, \tag{3.26}$$

and

$$\left(\frac{\partial u_0}{\partial x} + \frac{\partial v_0}{\partial y} \right) = 0. \tag{3.27}$$

This is called the *geostrophic relationship*. This relationship states that at the zeroth order the motion is balanced by the Coriolis force and the pressure gradient force.

The first order equations are found to be

$$\frac{\partial u_0}{\partial t} + u_0 \frac{\partial u_0}{\partial x} + v_0 \frac{\partial u_0}{\partial y} - v_1 - \left(\beta_0 y - \frac{\bar{\delta}}{2}y^2\right) v_0 = -\frac{\partial \psi_1}{\partial x}, \quad (3.28)$$

$$\frac{\partial v_0}{\partial t} + u_0 \frac{\partial v_0}{\partial x} + v_0 \frac{\partial v_0}{\partial y} + u_1 + \left(\beta_0 y - \frac{\bar{\delta}}{2}y^2\right) u_0 = -\frac{\partial \psi_1}{\partial y}, \quad (3.29)$$

and

$$F\left(\frac{\partial \psi_0}{\partial t} + u_0 \frac{\partial \psi_0}{\partial x} + v_0 \frac{\partial \psi_0}{\partial y}\right) - u_0 \frac{\partial \eta_B}{\partial x} - v_0 \frac{\partial \eta_B}{\partial y}$$

$$= \left(\frac{\partial u_1}{\partial x} + \frac{\partial v_1}{\partial y}\right) = 0. \quad (3.30)$$

Differentiating (3.29) with respect to x, substracting the differentiation of (3.28) with respect to y, and using (3.30), we obtain

$$\left(\frac{\partial}{\partial t} + u_0 \frac{\partial}{\partial x} + v_0 \frac{\partial}{\partial y}\right)(\nabla^2 \psi_0 - F\psi_0)$$

$$+ (\beta_0 - \bar{\delta}y + \beta_1)\frac{\partial \psi_0}{\partial x} - \beta_2 \frac{\partial \psi_0}{\partial y} = 0, \quad (3.31)$$

where

$$\beta_1 = \frac{\partial \eta_B}{\partial y}, \qquad \beta_2 = \frac{\partial \eta_B}{\partial x}. \quad (3.32)$$

The equation (3.31) is called the *potential vorticity equation*, and β_2 and β_1 are the topography slopes along the north–south and the west–east direction, respectively. This potential vorticity equation is a nonlinear equation. Linearizing the potential vorticity equation, we obtain,

$$\left(\frac{\partial}{\partial t} + U \frac{\partial}{\partial x} + V \frac{\partial}{\partial y}\right)(\nabla^2 \psi_0' - F\psi_0')$$

$$+ (B_1 + \beta_1)\frac{\partial \psi_0'}{\partial x} - (B_2 + \beta_2)\frac{\partial \psi_0'}{\partial y} = 0, \quad (3.33)$$

where $U(y,t)$, $V(x,t)$ are the eastward and the northward components of the basic flow, respectively,

$$B_1 = FU + \beta_0 - \bar{\delta}y - \frac{\partial^2 U}{\partial y^2}, \quad (3.34a)$$

and

$$B_2 = \frac{\partial^2 V}{\partial x^2} - FV. \quad (3.34b)$$

3.3   Equations Governing the Evolution of a Rossby Wave Packet

Here we simplify an individual single synoptic disturbance system as a single wave packet, as discussed in Chapter 2, particularly, the Rossby wave packet in the problem considered here. Then the geostrophic stream function for the synoptic disturbance can be considered a Rossby wave packet. Since β changes very slowly with latitude and, in most cases, the basic flows vary slowly with spatial variables and time in geophysical fluids, then the functions in our model (3.33) vary very slowly with spatial variables and time, as far as only large-scale topography with a smooth shape has been considered. However, the processes of the evolution of generated disturbance systems, for example, the slowly varying trough systems and cyclones after they are generated in real geophysical fluids, are carried on more slowly than are their generating processes. In such cases, the wave packet theory developed in Chapter 2 can be applied to investigate the present problem, since it possesses these characteristics. We restrict our discussion to the case in which the basic flows are varying slowly in spatial variables and time and in which the topographies are large scale with a smooth shape; namely, they change slowly in their spatial variables.

The time and space variables are introduced in the same manner as in Section 2.4:

$$T = \varepsilon t, \qquad X = \varepsilon x, \qquad Y = \varepsilon y \tag{3.35}$$

and

$$\psi_0' = \Psi(X, Y, T) e^{i\theta(X,Y,T)/\varepsilon}, \tag{3.36}$$

where

$$\Psi(X, Y, T) = \Psi_0(X, Y, T) + \varepsilon \Psi_1(X, Y, T) + \varepsilon^2 \Psi_2(X, Y, T) + \cdots, \tag{3.37}$$

and we define

$$\sigma = -\frac{\partial \theta}{\partial T}, \tag{3.38a}$$

$$m = \frac{\partial \theta}{\partial X}, \tag{3.38b}$$

and

$$n = \frac{\partial \theta}{\partial Y}, \tag{3.38c}$$

where σ, m, and n are called, respectively, the local frequency, the local wave number along the X-direction, and the local wave number along the Y-direction of the Rossby wave packet. And ε is a small parameter, which can also be defined as

$$\varepsilon \sim \frac{L}{a}, \tag{3.39}$$

where a again is the average radius of the earth. Therefore, the small parameter ε is really the ratio of the wave length scale of the wave packet and the wave length scale of the medium. The parameter ε defined here has the same order of magnitude as the Rossby number defined in (3.13) by the same symbol. Therefore, the small parameter introduced here can be considered as either one, and defined in (3.13) or (3.39).

After introducing the slowly variables, in order to balance the governing equation (3.33), we choose

$$\bar{\delta} \sim \varepsilon\delta_0, \tag{3.40}$$

where δ_0 is one order of magnitude. However, the order of magnitude of δ should be of the order of $O(\varepsilon^2)$, that is, one order less than that of β, which agrees with our scalings (3.8) in the middle latitudes.

Substituting (3.36) and (3.37) into the potential vorticity equation (3.33), in virtue of (3.35) and (3.38), we obtain the zeroth approximation, or the dispersion relation,

$$(\sigma - Um - Vn)K^2 + (B_1 + \beta_1)m - (B_2 + \beta_2)n = 0, \tag{3.41}$$

where

$$K^2 = m^2 + n^2 + F. \tag{3.42}$$

From the first approximation, the amplitude equation is

$$\left(\frac{\partial}{\partial T} + U\frac{\partial}{\partial X} + V\frac{\partial}{\partial Y}\right)K^2\Psi_0 - (\sigma - Um - Vn)$$
$$\times \left\{\left(\frac{\partial m}{\partial x} + \frac{\partial n}{\partial y}\right)\Psi_0 + 2\left(m\frac{\partial}{\partial X} + n\frac{\partial}{\partial Y}\right)\Psi_0\right\}$$
$$- \left\{(B_1 + \beta_1)\frac{\partial\Psi_0}{\partial X} - (B_2 + \beta_2)\frac{\partial\Psi_0}{\partial Y}\right\} = 0. \tag{3.43}$$

From the dispersion relation (3.41), the frequency equation is

$$\sigma = Um + Vn - \frac{m}{K^2}(B_1 + \beta_1) + \frac{n}{K^2}(B_2 + \beta_2). \tag{3.44}$$

Therefore, the phase velocities are

$$C_X = \frac{\sigma}{m} = U + V\frac{n}{m} - K^{-2}\left\{(B_1 + \beta_1) - \frac{n}{m}(B_2 + \beta_2)\right\} \tag{3.45}$$

and

$$C_Y = \frac{\sigma}{n} = \frac{m}{n}U + V - K^{-2}\left\{(B_1 + \beta_1)\frac{m}{n} - (B_2 + \beta_2)\right\}, \tag{3.46}$$

and the group velocities are

$$C_{gX} = \frac{\partial\sigma}{\partial m} = U - K^{-4}\{(B_1+\beta_1)K^2 - 2m[(B_1+\beta_1)m - (B_2+\beta_2)n]\} \tag{3.47}$$

and

$$C_{gY} = \frac{\partial \sigma}{\partial n} = V + K^{-4}\{(B_2+\beta_2)K^2 + 2n[(B_1+\beta_1)m - (B_2+\beta_2)n]\}. \quad (3.48)$$

Using (3.38), the definitions of σ, m, and n, (3.44), (3.47), and (3.48), we obtain

$$\frac{D_g\sigma}{DT} = m\left(\frac{\partial U}{\partial T} - K^{-2}\frac{\partial B_1}{\partial T}\right) + n\left(K^{-2}\frac{\partial B_2}{\partial T} + \frac{\partial V}{\partial T}\right), \quad (3.49)$$

$$\frac{D_g m}{DT} = -\left\{n\frac{\partial V}{\partial X} - \frac{m}{k^2}\frac{\partial \beta_1}{\partial X} + \frac{n}{K^2}\left(\frac{\partial B_2}{\partial X} + \frac{\partial \beta_2}{\partial X}\right)\right\}, \quad (3.50)$$

and

$$\frac{D_g n}{DT} = -\left\{m\frac{\partial U}{\partial Y} - \frac{m}{K^2}\left(\frac{\partial B_1}{\partial Y} + \frac{\partial \beta_1}{\partial Y}\right) + \frac{n}{K^2}\frac{\partial \beta_2}{\partial Y}\right\}, \quad (3.51)$$

in virtue of the theory in Chapter 2, namely (3.93a,b,c) or (2.96) to (2.98). In addition, the equations governing the whole structural change of the Rossby wave packet are derived as

$$\frac{D_g}{DT}(m^2 + n^2) = -2mn\left(\frac{\partial V}{\partial X} + \frac{\partial U}{\partial Y}\right) + \frac{2m^2}{K^2}\frac{\partial \beta_1}{\partial X}$$

$$- \frac{2mn}{K^2}\left(\frac{\partial B_2}{\partial X} + \frac{\partial \beta_2}{\partial X}\right) + \frac{2mn}{K^2}$$

$$\times \left(\frac{\partial B_1}{\partial Y} + \frac{\partial \beta_1}{\partial Y}\right) - \frac{2n^2}{K^2}\frac{\partial \beta_2}{\partial Y} \quad (3.52)$$

and

$$\frac{D_g}{DT}\left(-\frac{n}{m}\right) = \frac{\partial U}{\partial Y} - \frac{n^2}{m^2}\frac{\partial V}{\partial X} - K^{-2}\left(\frac{\partial B_1}{\partial Y} + \frac{\partial \beta_1}{\partial Y}\right)$$

$$+ \frac{n}{mK^2}\left(\frac{\partial \beta_2}{\partial Y} + \frac{\partial \beta_1}{\partial X}\right) - \frac{n}{m^2K^2}\left(\frac{\partial B_2}{\partial X} + \frac{\partial \beta_2}{\partial X}\right), \quad (3.53)$$

where

$$\frac{D_g}{DT} = \frac{\partial}{\partial T} + C_{gX}\frac{\partial}{\partial X} + C_{gY}\frac{\partial}{\partial Y}, \quad (3.54)$$

which is the material derivative along the group velocity of the wave packet and is the same as (2.92) or (2.99a). Therefore, all the equations governing the evolution of the Rossby wave packet (3.47) to (3.54) have been obtained.

Let

$$\Psi_0 = |\Psi_0|e^{i\alpha(X,Y,T)}, \quad (3.55)$$

where

$$|\Psi_0| = |\Psi_0(X,Y,T)|. \quad (3.56)$$

Using the above relations, we rewrite eq. (3.43) as

$$\frac{D_g \alpha}{DT} = 0,$$ (3.57)

$$\left(\frac{\partial}{\partial T} + U \frac{\partial}{\partial X} + V \frac{\partial}{\partial Y} \right) K^2 |\Psi_0| + K^{-2} \{ (B_1 + \beta_1) m$$

$$- (B_2 + \beta_2) n \} \left\{ \left(\frac{\partial m}{\partial X} + \frac{\partial n}{\partial Y} \right) |\Psi_0| + 2 \left(m \frac{\partial}{\partial X} + n \frac{\partial}{\partial Y} \right) |\Psi_0| \right\}$$

$$- \left\{ (B_1 + \beta_1) \frac{\partial}{\partial X} - (B_2 + \beta_2) \frac{\partial}{\partial Y} \right\} |\Psi_0| = 0,$$ (3.58)

by separating it into its real and imaginary parts. Equation (3.57) states that the wave packet is propagated along with the group velocity, whereas eq. (3.58) is the amplitude equation of the wave packet.

3.4 Integral Properties of Wave Packets: Barotropic Instability Theorems

3.4.1 IN THE ZONAL BASIC FLOW

Suppose the basic flow is zonal and the topography is oriented east–westward. Given these assumptions, the governing equations (3.49) to (3.53) become

$$\frac{D_g \sigma}{DT} = - \left(m \frac{\partial U}{\partial T} - \frac{m}{K^2} \frac{\partial B}{\partial T} \right),$$ (3.59)

$$\frac{D_g m}{DT} = 0,$$ (3.60)

$$\frac{D_g n}{DT} = - \left(m \frac{\partial U}{\partial Y} - \frac{m}{K^2} \frac{\partial B}{\partial Y} \right),$$ (3.61)

$$\frac{D_g}{DT} (m^2 + n^2) = -2mn \frac{\partial U}{\partial Y} + \frac{2mn}{K^2} \frac{\partial B}{\partial Y},$$ (3.62)

and

$$\frac{D_g}{DT} \left(-\frac{n}{m} \right) = \frac{\partial U}{\partial Y} - K^{-2} \frac{\partial B}{\partial Y},$$ (3.63)

where

$$B = B_1 + \beta_1.$$ (3.64)

The amplitude equation (3.58) becomes

$$\left(\frac{\partial}{\partial T} + U \frac{\partial}{\partial X} \right) K^2 |\Psi_0| + \frac{mB}{K^2} \left\{ \left(\frac{\partial m}{\partial X} + \frac{\partial n}{\partial Y} \right) |\Psi_0| \right.$$

$$\left. + 2 \left(m \frac{\partial}{\partial X} + n \frac{\partial}{\partial Y} \right) |\Psi_0| \right\} - B \frac{\partial}{\partial X} |\Psi_0| = 0.$$ (3.65)

These equations are similar to those obtained by Zeng (1982).

Equation (3.65) can be rewritten as

$$\frac{D_g}{DT}\left(\frac{K^2}{2}|\Psi_0|^2\right) + \frac{K^2}{2}|\Psi_0|^2\nabla\cdot\mathbf{C}_g = |\Psi_0|^2 mn\frac{\partial U}{\partial Y}. \tag{3.66}$$

Using (3.62), we obtain the following integral properties of wave packet from (3.66):

$$\frac{\partial}{\partial T}\int\int_{(S)}\frac{K^2}{2}|\Psi_0|^2 dXdY = \int\int_{(S)} mn\frac{\partial U}{\partial Y}|\Psi_0|^2 dXdY, \tag{3.67}$$

$$\int\int_{(S)} B^{-1}\frac{\partial}{\partial T}(K^4|\Psi_0|^2)dXdY = 0, \tag{3.68}$$

and

$$\int\int_{(S)}\left\{\frac{\partial}{\partial T}\left(\frac{K^2}{2}|\Psi_0|^2\right) + \frac{U_r - U}{B}\frac{\partial}{\partial T}\left(\frac{K^4}{2}|\Psi_0|^2\right)\right\}dXdY = 0, \tag{3.69}$$

where S denotes the whole region occupied by the wave packet. Hence U_r is an arbitrary constant, and $(K^2/2|\Psi_0|^2)$ and $(K^4/2)|\Psi_0|^2)$ are the analogues of the density of the wave packet energy and the enstrophy, respectively. The total energy of the wave packet is given by

$$E = \int\int_{(S)}\frac{K^2}{2}|\Psi_0|^2 dXdY, \tag{3.70}$$

so that (3.67) describes the change of total energy of the wave packet with time, whereas (3.68) is the conservation of total wave-action and (3.69) is the conservation of enstrophy. The detailed derivations of these integral properties can be found in Appendix A.

If the wave packet is said to be unstable when its energy is increased with time, and stable when its energy is decreased with time, from the above integral properties, we can obtain the following barotropic instability theorems.

Theorem I. *In a basic zonal flow, such as* $(\partial U/\partial Y) > 0$, *the west-tilting wave packet will increases its energy with time and become unstable, whereas the east-tilting wave packet will decreases its energy and become stable. In a basic zonal flow such as* $(\partial U/\partial Y) < 0$, *the west-tilting wave packet will decrease its energy and become stable, whereas the east-tilting wave packet will increase its energy and become unstable. The definition and discussion of the west-tilting and the east-tilting wave packet can be found in Section 3.5.1.*

Theorem II. *The necessary condition for barotropic instability of the wave packet is that there is at least one point (X_j, Y_j) over the whole domain of the wave packet at which $B = 0$, that is,*

$$B \equiv FU + \beta_0 - \delta_0 Y - \frac{\partial^2 U}{\partial y^2} = 0; \qquad at \ X = X_j, Y = Y_j. \qquad (3.71)$$

Theorem III. *The necessary condition for barotropic instability of the wave packet is that there is at least some place over the domain of the wave packet where*

$$\frac{U_r - U}{B} < 0. \qquad (3.72)$$

Theorem I is the consequence of the first integral property of the wave packet (3.67). It is essentially the same as the previous finding of barotropic instability associated with the wave-mean interaction (e.g., Pedlosky 1987).

Theorem II is from the second integral property of the wave packet (3.68). It reduces to the Kuo's (1949) theorem when there is no topography, and no free surface. In addition, if there is no rotation, it is further reduced to the Rayleigh theorem, namely, the Rayleigh inflexion-point theorem (Drazin and Reid, 1981).

Similarly, Theorem III is from the third integral property of the wave packet (3.69). It is reduced to the Fjørtoft theorem (Drazin and Reid, 1981) when there is no topography, and no free surface.

However, the question is, "Is it necessary that a barotropically unstable wave packet always remain unstable?" These theorems cannot answer this question. This question, nevertheless, is fundamental and of great importance in theoretical studies and practical work. The answer, in general, is negative. Then, how will the wave packet change its structure, and is it possible that the wave packet undoes its unstable structure and assumes a stable structure, and how does it do it? We try to answer these questions in the next section. The structure of developing and decaying wave packet is discussed in Chapter 7.

If the energy conservation (3.67) and the total wave number equation (3.62) are carefully compared, it is found that there is a relationship between the change in the total wave number and the change in the total energy of the wave packet. The equation of change in the total wave number (3.62) can be rewritten in the zonal basic flow as follows:

$$\frac{D_g}{DT}(m^2 + n^2) = -\frac{2mn}{K^2} \frac{\partial U}{\partial Y}(m^2 + n^2). \qquad (3.73)$$

Now, it is easily seen that if the total wave number is increased, then the total wave energy decreases. Though the total wave number is decreased, the wave packet increases its total energy. Since the wave length scale is inversely proportional to the total wave number, the relationship can be stated as follows: the wave packet's energy increases (decreases) as the wave length scale of the wave packet increases (decreases). Therefore, the structural change in the wave packet is consistent with the change in the total energy. Moreover, the structural change can describe the change in the total energy. We discuss the structural change of the wave packet in Section 3.5.

3.4.2   IN THE ASYMMETRIC BASIC FLOW

We consider the topography and asymmetric basic flow as satisfying the following conditions:

$$B_1 + \beta_2 = B(X, Y, T) \tag{3.74a}$$

and

$$B_2 + \beta_2 = 0. \tag{3.74b}$$

Then, the group velocities (3.47) and (3.48) become

$$C_{gX} = U - \frac{B(K^2 - 2m^2)}{K^4} \tag{3.75}$$

and

$$C_{gY} = V + \frac{2Bmn}{K^4}. \tag{3.76}$$

The equation governing the whole wave number of the wave packet (3.52) can be written as

$$\begin{aligned}
\frac{D_g K^2}{DT} = &\frac{2m}{K^2}\left(m\frac{\partial B}{\partial X} + n\frac{\partial B}{\partial Y} \right) \\
&- 2mn\left(\frac{\partial V}{\partial X} + \frac{\partial U}{\partial Y} \right) \\
&- 2\left(m^2\frac{\partial U}{\partial X} + n^2\frac{\partial V}{\partial Y} \right),
\end{aligned} \tag{3.77}$$

where

$$K^2 = m^2 + n^2 + F \tag{3.78}$$

and

$$\frac{D_g}{DT} = \frac{\partial}{\partial T} + C_{gX}\frac{\partial}{\partial X} + C_{gY}\frac{\partial}{\partial Y}, \tag{3.79}$$

where F, the Froude number, is constant.

In the present case, the evolution equation of the amplitude of the wave packet (3.58) becomes

$$\left(\frac{\partial}{\partial T} + U\frac{\partial}{\partial X} + V\frac{\partial}{\partial Y}\right)K^2|\Psi_0| + \frac{mB}{K^2}\left[\left(\frac{\partial m}{\partial X} + \frac{\partial n}{\partial Y}\right)|\Psi_0|\right.$$

$$\left. +2\left(m\frac{\partial|\Psi_0|}{\partial X} + n\frac{\partial|\Psi_0|}{\partial Y}\right)\right] - B\frac{\partial|\Psi_0|}{\partial X} = 0. \tag{3.80}$$

This equation also can be rewritten as

$$K^2\frac{D_g|\Psi_0|}{DT} + |\Psi_0|\left[\frac{D_g K^2}{DT} - \left(\frac{2Bm^2}{K^4} - \frac{B}{K^4}\right)\frac{\partial K^2}{\partial X}\right.$$

$$\left. - \frac{2Bmn}{K^4}\frac{\partial K^2}{\partial Y}\right] + \frac{Bm|\Psi_0|}{K^2}\left(\frac{\partial m}{\partial X} + \frac{\partial n}{\partial Y}\right) = 0. \tag{3.81}$$

In virtue of (3.75) to (3.77), the above equation can be further written as

$$K^2\frac{D_g|\Psi_0|}{DT} + \frac{1}{2}|\Psi_0|\frac{D_g K^2}{DT} + \frac{K^2|\Psi_0|}{2}\nabla\cdot\mathbf{C}_g - mn|\Psi_0|\left(\frac{\partial U}{\partial Y} + \frac{\partial V}{\partial X}\right)$$

$$+ \frac{|\Psi_0|}{2}\frac{\partial B}{\partial X} - |\Psi_0|\left[\frac{K^2}{2}\left(\frac{\partial U}{\partial X} + \frac{\partial V}{\partial Y}\right)\right.$$

$$\left. + m^2\frac{\partial U}{\partial X} + n^2\frac{\partial V}{\partial Y}\right] = 0. \tag{3.82}$$

Multiplying (3.82) by $|\Psi_0|$ and integrating over the whole area of the wave packet, we obtain the following integral relation of change in wave packet energy, namely,

$$\frac{\partial}{\partial T}\int\int_S \frac{1}{2}K^2|\Psi_0|^2 dXdY = \int\int_S |\Psi_0|^2 mn\left(\frac{\partial U}{\partial Y} + \frac{\partial V}{\partial X}\right)dXdY$$

$$- \int\int_S \frac{|\Psi_0|^2}{2}\frac{\partial B}{\partial X}dXdY$$

$$+ \int\int_S \frac{|\Psi_0|^2}{2}K^2\left(\frac{\partial U}{\partial X} + \frac{\partial V}{\partial Y}\right)dXdY$$

$$+ \int\int_S |\Psi_0|^2\left(m^2\frac{\partial U}{\partial X} + n^2\frac{\partial V}{\partial Y}\right)dXdY. \tag{3.83}$$

This equation is obviously reduced to (3.67) when B does not vary with X and there is no meridional component of basic flow. Therefore, in this case, the change in the wave packet energy depends on the basic flow in a rather complicated manner, as governed by (3.83).

Multiplying (3.81) by $2K^2|\Psi_0|/B$, and integrating over the whole area of the wave packet, we derive the integral relation of change in the analogues of the density of the wave packet enstrophy as follows:

$$\int\int_S \frac{1}{B}\frac{\partial}{\partial T}(K^2|\Psi_0|^2)^2 dXdY = -\int\int_S \frac{U}{B}\frac{\partial(|\Psi_0|^2 K^4)}{\partial X}dXdY$$
$$-\int\int_S \frac{V}{B}\frac{\partial(|\Psi_0|^2 K^4)}{\partial Y}$$
$$+\int\int_S n^2\frac{\partial|\Psi_0|^2}{\partial X}dXdY$$
$$-2\int\int_S m\frac{\partial n|\Psi_0|^2}{\partial Y}dXdY. \quad (3.84)$$

Now it is obvious that the enstrophy in the present case is no longer conserved. However, (3.84) will be reduced to (3.68) when B is independent on X and there is no meridional component of basic flow. This statement can be proven in the following way. Since B is independent on X, so is U. The first term on the right-hand side of (3.84) can be written

$$\int\int_S \frac{\partial}{\partial X}\left(\frac{U}{B}|\Psi_0|^2 K^4\right)dXdY,$$

which becomes zero when the appropriate boundary condition is applied. The second term on the right-hand side of (3.84) is zero due to the absence of the meridional component of basic flow. The third term and the fourth term on the right-hand side of (3.84) can be written as

$$\int\int_S \left(n^2\frac{\partial|\Psi_0|^2}{\partial X} - 2m\frac{\partial n|\Psi_0|^2}{\partial Y}\right)dXdY$$
$$=\int\int_S \left[\frac{\partial n^2|\Psi_0|^2}{\partial X} - 2|\Psi_0|^2 n\frac{\partial n}{\partial X} - 2\frac{\partial(mn|\Psi_0|^2)}{\partial Y} + n|\Psi_0|^2\frac{\partial m}{\partial Y}\right]dXdY,$$
$$= 2\int\int_S n|\Psi_0|^2\left(\frac{\partial m}{\partial Y} - \frac{\partial n}{\partial X}\right)dXdY = 0,$$

in virtue of the definitions of m and n, that is, (2.63).

Thus, indeed, the present results are the extensions of the previous ones. Similar integral properties were first derived by Yang and Yang (1988, 1990) in the investigation of wave packets and teleconnections, which are discussed in Section 2 of Chapter 8.

3.5   Structural Change of a Rossby Wave Packet

Previous studies have mainly focused on the effects of basic flows, topography, and variation of the Coriolis parameter with latitude upon the propagation properties in Rossby waves (e.g., Phillips, 1965; Platzman, 1968; Pedlosky, 1987; Hoskins and Karoly, 1982). However, these propagation properties cannot completely describe the evolution of the packet. Does the Rossby wave packet change its structure with time due to these factors? If it does, then how do the factors alter the structure? We dicuss the results in the following paragraphs.

If the governing equations (3.47) to (3.53) are carefully examined, one easily finds that if the basic flow and the slopes of topography are linear function of space, then the equations governing the local wave number can be decoupled from the equations governing the path of the wave packet mathematically, as the result of the structure independence theorem discussed in Section 2.4. Since the wave packet is propagated along the group velocity, the material derivative along the group velocity can be treated as an ordinary material derivative. The system of the partial differential equations consisting of the equations governing the local wave number can be treated as if it were an ordinary differential system, as discussed in Section 2.4.

3.5.1 THE EARTH'S ROTATION: THE δ-EFFECT

The δ-effect was first introduced by Yang (1987, 1988a); it refers to the effect of the second derivative of the Coriolis parameter, with respect to latitude, on the structural change of the Rossby wave packet. If we only consider the δ-effect, for instance, modeling conditions near the polar regions where $\beta = 0$, then (3.50) to (3.53) will become

$$\frac{D_g m}{DT} = 0, \tag{3.85}$$

$$\frac{D_g n}{DT} = -\frac{m}{K^2}\delta_0, \tag{3.86}$$

$$\frac{D_g}{DT}(m^2 + n^2) = -\frac{2mn}{K^2}\delta_0, \tag{3.87}$$

and

$$\frac{D_g}{DT}\left(-\frac{n}{m}\right) = \frac{\delta_0}{K^2}, \tag{3.88}$$

where δ_0 is always positive in the Northern Hemisphere and always negative in the Southern Hemisphere. In the following, we will only consider δ to be positive, (thus only the Northern Hemisphere is considered). The results can easily be extended to the Southern Hemisphere.

From the definition of the local wave number when the phase function is constant, the slope of the trough (or ridge) line of the wave packet is derived as

$$\left(\frac{\partial X}{\partial Y}\right)\Big|_{\theta=\text{constant}} = -\frac{\partial\theta/\partial Y}{\partial\theta/\partial X} = -\frac{n}{m}. \tag{3.89}$$

Therefore, $-n/m > 0$ means that the slope of the disturbance system is tilted to the west with respect to the Y-axis, and $-n/m < 0$ means that

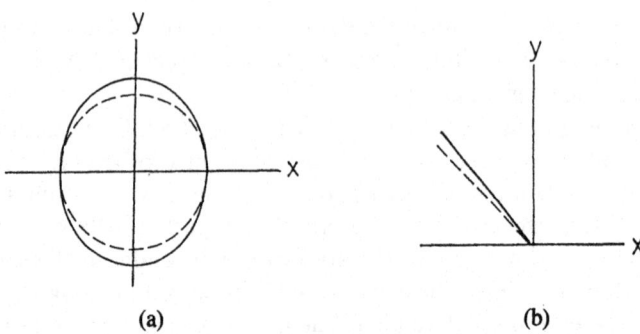

FIGURE 3.2. The δ-effect. (a) Change of the spatial scale and (b) change of the trough (or ridge) line of the Rossby wave packet due to the second derivative with latitude. Key: - - -, at $T = 0$; —, at $T = 200$. After Yang (1987).

the disturbance system is tilted to the east. Without a loss of generality, in this chapter, we suppose that $m > 0$. Then, the disturbance system has a westward or eastward tilt with respect to the Y-axis for $n > 0$ or $n < 0$. We use the terminology *trough* (or *ridge*) *line* instead of the slope of the system, which is more commonly used in the atmospheric sciences, below.

From (3.85) to (3.88), we find that δ_0 will not affect the packet's longitudinal scale but that it does affect its latitudinal scale. This enlarges the packet's latitudinal scale by decreasing the latitudinal wave number. (Here, and hereafter in this chapter, the spatial scale change will always be considered in a westward tilting system, that is, $n > 0$, unless stated otherwise. Obviously, the eastward tilting disturbances can also be discussed using the same method.) This means that the whole spatial scale will become enlarged. Equation (3.88) indicates that the δ-effect makes the westward tilting trough move toward the Y-axis, whereas it makes the eastward tilting trough line move eastward. The reason is very easy to understand, viz., the northern part of the Rossby wave packet will move eastward faster than the southern part, due to the variation of the change in the β with respect to latitude, that is, δ. This phenomenon is well known in geophysical fluids. However, the first rigorous mathematical proof for the phenomenon was only given recently (Yang, 1987).

For example, when we take $\delta_0 = 0.5$, which corresponds to the case in which the wave length is about 4000 km in the middle latitude regions, we compute the δ-effect, which is shown in Figure 3.2. In Figure 3.2a, the dashed line is an ideal streamline at $T = 0$, and the solid line is the streamline after 200 dimensionless time steps, which is about five days in the dimensional time scale, a typical time scale for synoptic systems. Hereafter in this chapter, we will always take $F = 1$ and $\varepsilon = 0.1$ and choose $m = n = 6$ at $T = 0$, in all our model computations. For convenience of comparison, we take an ideal streamline at $T = 0$ as the isolated symmetric eddy. In Figure 3.2b, the dashed line is the trough line at $T = 0$; the solid line is the trough line after 200 dimensionless time steps. The results show that the latitudinal scale of the wave packet increases by 20%, so that the whole spatial scale has been increased by about 8% and the trough line tilts toward the Y-axis by 5°.

3.5.2 ASYMMETRIC BASIC FLOWS

If only the main part of the asymmetric basic flow effect on the structural change of the Rossby wave packet is considered, then (3.50) to (3.53) become

$$\frac{D_g m}{DT} = -\frac{n}{K^2}(m^2 + n^2)\frac{\partial V}{\partial X}, \tag{3.90}$$

$$\frac{D_g n}{DT} = -\frac{m}{K^2}(m^2 + n^2)\frac{\partial U}{\partial Y}, \tag{3.91}$$

$$\frac{D_g}{DT}(m^2 + n^2) = -2mn\frac{m^2 + n^2}{K^2}\left(\frac{\partial U}{\partial Y} + \frac{\partial V}{\partial X}\right), \tag{3.92}$$

and

$$\frac{D_g}{DT}\left(-\frac{n}{m}\right) = \frac{m^2 + n^2}{K^2}\left(\frac{\partial U}{\partial Y} - \frac{n^2}{m^2}\frac{\partial V}{\partial X}\right). \tag{3.93}$$

As examples, we consider two typical asymmetric basic flows, that often occur in southeast Asia, that is, the southwesterly jet and the southeasterly jet. Figure 3.3a and Figure 3.3b show an ideal southwesterly jet and an ideal southeasterly jet, respectively.

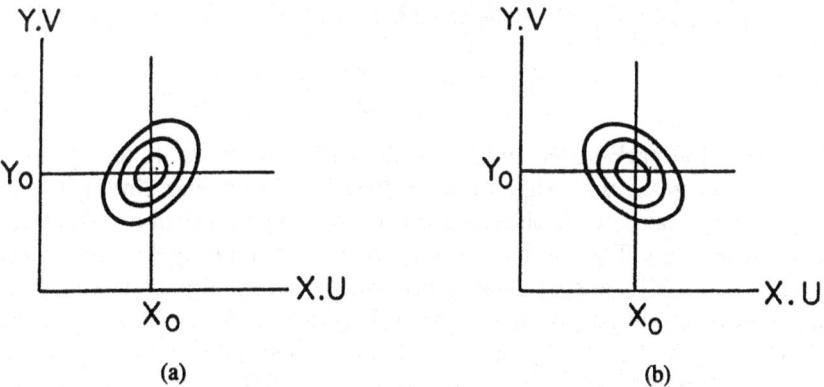

(a)   (b)

FIGURE 3.3. Ideal jet profiles. (a) The southwesterly jet and (b) the southeasterly jet. The centers are located at (X_0, Y_0), and the basic isotachs are indicated by solid lines. X and Y are eastward and northward, respectively.

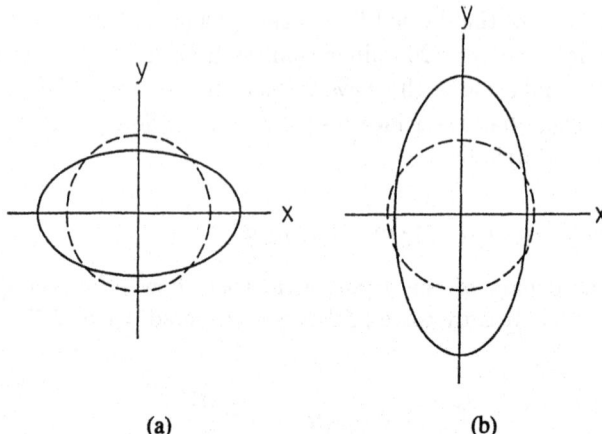

(a)   (b)

FIGURE 3.4. The spatial scale change of the Rossby wave packet in the south-westerly jet. (a) On the left side and (b) on the right side of the jet. Key: - - -, an ideal streamline at $T = 0$; —, the ideal streamline at $T = 200$. After Yang (1987).

On the left-hand side of a southwesterly jet, one has

$$\frac{\partial U}{\partial Y} < 0, \qquad \frac{\partial V}{\partial X} > 0. \tag{3.94}$$

Thus, from (3.90) to (3.93), one could know that at such a location the longitudinal scale of the packet will increase, while its latitudinal scale will decrease due to the influence of the southwesterly jet. Moreover, the westward tilting trough line will become more westwardly, while the eastward tilting trough line will tend toward the Y-axis.

On the right-hand side of the southwesterly jet,

$$\frac{\partial U}{\partial Y} > 0, \qquad \frac{\partial V}{\partial X} < 0. \tag{3.95}$$

Therefore, the packet's latitudinal scale will increase and its longitudinal scale will decrease with time. The westward tilting trough line will tend towards the Y-axis, while the eastward tilting trough line will become more eastwardly tilted. The southeasterly jet would act similarly, but in reverse.

The main results of spatial structural changes of the Rossby wave packet in a southwesterly jet are illustrated in Figure 3.3. In this figure, we take $\partial U/\partial Y = -0.02$ and $\partial V/\partial X = 0.01$ on the left side of the jet, which corresponds to the basic flow changes of about 2 m/sec in the U component and 1 m/sec in the V component 100 km out from the jet carrier. The dashed line is an ideal streamline of the packet at $T = 0$, and the solid line is the streamline after 200 dimensionless time steps. The results demonstrates that in such a southwesterly jet, the packet's longitudinal scale enlarges

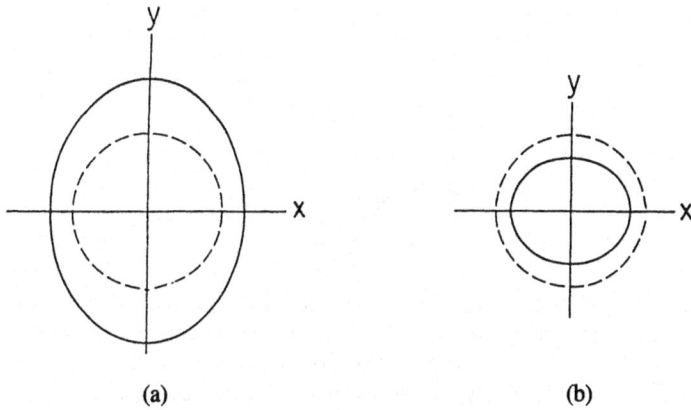

(a)   (b)

FIGURE 3.5. The spatial scale change of the Rossby wave packet in the south-easterly jet. (a) On the left side and (b) on the right side of the jet. Key: - - -, an ideal streamline at $T = 0$; —, the ideal streamline at $T = 200$. After Yang (1987).

40%, while its latitudinal scale shrinks 20%. In addition, the whole spatial scale has been decreased by about 10%, and its westward tilting trough line tends toward the west by about 16° on the left side of the jet. However, on the right side of such a southwesterly jet, where we chose $\partial U/\partial Y = 0.02$ and $\partial V/\partial X = -0.01$, the results further show that the packet's longitudinal scale decreases 10% and its latitudinal scale increases 80%. Hence, the whole spatial scale has been increased by about 12%, and the westward tilting trough line tends toward the Y-axis about 18° (see Figure 3.4).

Figure 3.5 shows the results of an ideal southeasterly jet. Here again, we have computed the results after 200 dimensionless time steps in considering such a southeasterly jet, where $\partial U/\partial Y = 0.02$ and $\partial V/\partial X = 0.01$. The results show that, on the left side of the jet, the whole scale will increase, while the westward tilting trough line will tend toward the Y-axis. The longitudinal scale increases about 30%, while the latitudinal scale increases 70% and the trough line about 8%. However, on the right side of the jet, the whole scale of the packet will decrease, while the westward tilting trough line will tilt toward the west. The longitudinal scale shrinks about 20%, the latitudinal scale about 30%. Therefore, the whole spatial scale has been decreased about 35%, and the trough line tilts westward about 4° after 200 dimensionless time steps. Figure 3.6 shows the results of a trough line, where the dashed line is the trough line at $T = 0$, and the solid lines A, B, C, and D are the westward tilting trough line after 200 dimensionless time steps, corresponding to the cases in which the wave packet is on the left side and the right side of a southwesterly jet and on the left side and the right side of a southeasterly jet, respectively.

It can be seen from these figures that there is no obvious difference be-

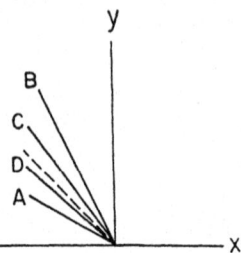

FIGURE 3.6. The trough (or ridge) lines. The dashed line is $T = 0$; solid lines A, B, C, and D are trough lines at $T = 200$, corresponding to the cases on the left side and the right side of the southwesterly jet and on the left side and the right side of the southeasterly jet, respectively. After Yang (1987).

tween the effects of these two jets upon the packet. In the southwesterly jet, the change of the packet's whole spatial scale depends on the magnitudes of $\partial U/\partial Y$ and $\partial V/\partial X$, since its latitudinal scale decreases as its longitudinal scale increases. But in a southeasterly jet, both the longitudinal scale and the latitudinal scale increase or decrease (depending on the location relative to the jet) simultaneously. Therefore, the change of the spatial scale in the southeasterly jet will be significant in some cases, that is, the jet will largely affect the spatial scale of the packet. Furthermore, from the above analysis (Figure 3.4) one can find that the Rossby wave packet, when located on the left side of the southwesterly jet, will enlarge its longitudinal scale and shrink its latitudinal scale so that finally the wave packet easily can be absorbed by the zonal flow. This phenomenon was called *rotational adaptation* by Zeng (1979). However, here the rotational adaptation is caused not only by the rotation, but rather by the joint effect of the rotation, the topography, and the jet. Yet, when located at the right side of southwesterly jet, rotational adaptation does not occur. If the enlargement of the longitudinal scale is simply considered as developing the disturbance system, or taking the latitudinal scale to be infinite, that is, $n = 0$, the result suggests that the disturbance develops more easily on the right side of the southwesterly jet than on the left.

3.5.3 EFFECTS OF TOPOGRAPHY

When the effect of topography is considered, eqs. (3.50) to (3.53) become

$$\frac{D_g m}{DT} = \frac{1}{K^2}\left(m\frac{\partial\beta_1}{\partial X} - n\frac{\partial\beta_2}{\partial X}\right), \tag{3.96}$$

$$\frac{D_g n}{DT} = \frac{1}{K^2}\left(m\frac{\partial\beta_1}{\partial Y} - n\frac{\partial\beta_2}{\partial Y}\right), \tag{3.97}$$

$$\frac{D_g}{DT}(m^2 + n^2) = \frac{2}{K^2}\left\{m^2\frac{\partial\beta_1}{\partial X} - n^2\frac{\partial\beta_2}{\partial Y} + mn\left(\frac{\partial\beta_1}{\partial Y} - \frac{\partial\beta_2}{\partial X}\right)\right\}, \tag{3.98}$$

and

$$\frac{D_g}{DT}\left(-\frac{n}{m}\right) = \frac{1}{K^2}\left(-\frac{\partial\beta_1}{\partial Y} + \frac{n}{m}\frac{\partial\beta_1}{\partial X} - \frac{n^2}{m^2}\frac{\partial\beta_2}{\partial X} + \frac{n}{m}\frac{\partial\beta_2}{\partial Y}\right). \quad (3.99)$$

(1) Linearly Distributed Topography

Equations (3.96) to (3.99) show that the linearly distributed topography, that is, the height of the topography, is a linear function of X and Y and has no effect on the structural change of the Rossby wave packet. In other words, the packet above the linearly sloping topography can have the same structure as a packet above a flat surface. From previous studies (e.g., Pedlosky, 1987) or eqs. (3.45) to (3.48) in Section 3.3, it is known that the linearly sloping topography does affect the properties of the propagations in both phase and energy. One interpretation is that when the wave packet is propagating in the linearly sloping topography, it cannot acquire or lose its potential vorticity from or into the environmental vorticity field with higher or lower speed. This is due to the linearly sloping topography, which can only produce a uniform environmental vorticity gradient field. Therefore, the linearly sloping topography cannot change the structure of the wave packet. This is reminiscent of the β-plane approximation. In that case, the Rossby wave can be found with its phase velocity and energy propagation. However, we did not find a change in the structure of the Rossby wave. The reason is that, in the potential vorticity equation, on the earth's β-plane the gradient of planetary vorticity was artificially taken to be constant. So, after arriving at its new position the Rossby wave could not gain either more or less at the environment planetary vorticity that is required in order to change the wave packet structure.

(2) Nonlinearly Distributed Topography

Nonlinear topography, that is, topography that is described by a nonlinear function of X and Y, however, has a substantially different effect on the structure of the Rossby wave packet than has linear topography. Nonlinear topography will affect not only the phase velocity and the energy propagation properties but also the packet's structure. For simplicity, four types of topographies are considered. Other more complex types can be discussed by using the same method.

North–South-Oriented Topography. The Rocky Mountains may be considered a typical example. In this case,

$$\beta_1 = 0, \qquad \frac{\partial\beta_2}{\partial Y} = 0, \qquad \frac{\partial\beta_2}{\partial X} = -a < 0, \qquad (3.100)$$

where a is a positive constant. Then, eqs. (3.96) to (3.99) can be written

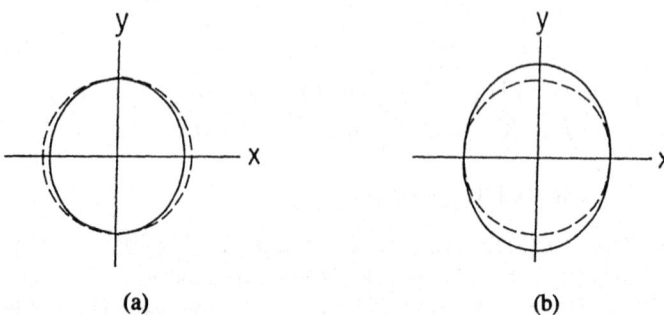

FIGURE 3.7. The spatial scale change of the Rossby wave packet on (a) the north–south- and (b) the east–west-oriented topography. Key: - - -, an ideal streamline at $T = 0$; —, the ideal streamline at $T = 200$.

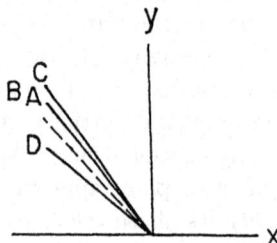

FIGURE 3.8. The trough (or ridge) lines. The dashed line is at $T = 0$; solid lines A, B, C, and D are at $T = 200$, corresponding to the north–south-oriented, the east–west-oriented, convex and concave topography, respectively, where B coincides with A.

as follows:

$$\frac{D_g m}{DT} = \frac{na}{K^2}, \tag{3.101}$$

$$\frac{D_g n}{DT} = 0, \tag{3.102}$$

$$\frac{D_g}{DT}(m^2 + n^2) = \frac{2mna}{K^2}, \tag{3.103}$$

and

$$\frac{D_g}{DT}\left(-\frac{n}{m}\right) = -\frac{n^2 a}{m^2}. \tag{3.104}$$

Obviously, in this case, the Rossby wave packet shrinks its longitudinal scale while its latitudinal scale remains unaltered. The whole spatial scale shrinks with time. The westward tilting trough tilts more toward the Y-axis, while the eastward tilting trough line tilts further eastward. Figure 3.7a and Figure 3.8 show the results of such a kind of topography as $a = 0.8$, where the dashed line is the ideal streamline at $T = 0$, while the solid line is the streamline after 200 dimensionless time steps. Here, if we consider the topography to be a hyperbolic surface, then $a = 0.8$ corresponds to $\eta_B = H - 0.8xX$. One can find from the figure that its latitudinal scale remains the same, while its longitudinal scale shrinks 10%, which implies that the whole spatial scale has shrunk about 10%, and its westward tilting trough line tends to the Y-axis about 5°, after 200 dimensionless time steps, as shown in Figure 3.7a and Figure 3.8 by the solid line A.

East–West-Oriented Topography. Here,

$$\frac{\partial \beta_1}{\partial Y} = -b < 0, \qquad \frac{\partial \beta_1}{\partial X} = 0, \qquad \beta_2 = 0, \tag{3.105}$$

where b is a positive constant. Thus, eqs. (3.96) to (3.99) will take the following forms:

$$\frac{D_g m}{DT} = 0, \tag{3.106}$$

$$\frac{D_g n}{DT} = -\frac{mb}{K^2}, \tag{3.107}$$

$$\frac{D_g}{DT}(m^2 + n^2) = -\frac{2mnb}{K^2}, \tag{3.108}$$

and

$$\frac{D_g}{DT}\left(-\frac{n}{m}\right) = \frac{b}{K^2}. \tag{3.109}$$

Therefore, an east–west-oriented topography will not alter the Rossby wave packet's longitudinal scale but will enlarge the entire spatial scale. The westward tilting trough line will tend toward the Y-axis, while the eastward tilting trough line will tilt further east. Figure 3.7b shows the result of this kind of topography, when $b = 0.5$, corresponding to $\eta_B = H - 0.5yY$, where again the dashed line is an ideal streamline at $T = 0$ and the solid line is the streamline after 200 dimensionless time steps. It is shown that after 200 dimensionless time steps, the packet's longitudinal scale remains the same as before, while its latitudinal scale will enlarge about 20%. In this case, the whole spatial scale has been enlarged about 8%, and the packet's westward tilting trough line will tend toward the Y-axis by about 5°, which is the solid line B and which coincides with the solid line A in Figure 3.8.

Convex Topography. The Tibet Plateau may be considered an example of this type of topography. Then,

$$\frac{\partial \beta_1}{\partial X} = 0, \qquad \frac{\partial \beta_1}{\partial Y} = -b < 0, \qquad \frac{\partial \beta_2}{\partial X} = -a < 0, \qquad \frac{\partial \beta_2}{\partial Y} = 0, \quad (3.110)$$

where a and b are taken as positive constants. The governing equations are

$$\frac{D_g m}{DT} = \frac{na}{K^2},$$ (3.111)

$$\frac{D_g n}{DT} = -\frac{mb}{K^2},$$ (3.112)

$$\frac{D_g}{DT}(m^2 + n^2) = \frac{2mn(a-b)}{K^2},$$ (3.113)

and

$$\frac{D_g}{DT}\left(-\frac{n}{m}\right) = \frac{m^2 b + n^2 a}{m^2 K^2}.$$ (3.114)

From these equations, it can be seen that the convex topography will shrink the longitudinal scale. If we consider an X-propagating disturbance system, the result strongly suggests that the mountains, especially the Rocky Mountains (where $b = 0$), will decrease the longitudinal component of the system. This result is consistent with that of the spectral general circulation models of Hayashi and Golder (1983). They found that, in the absence of mountains, the length scales of the eastward moving wave (wave numbers 4 to 6) are markedly increased in the Northern Hemisphere. The mountains did decrease the eastward moving wave components. The change of the whole spatial scale depends on the magnitudes of a and b. When $a = b$, which describes a symmetric topography about the vertical axis, the whole scale will not change. The westward tilting trough line will tend toward the Y-axis, while the eastward tilting trough line will be tilted more toward the east. If we consider the Tibet Plateau to be a topography typical of the case $b > a$, the topography enlarges the packet's whole spatial scale and moves its westward tilting trough line toward the Y-axis, while its eastward tilting trough line will tilt more to the east. These results are consistent with the observations and the results of Hayashi and Golder (1983). This kind of topography has the same qualitative effect on the structural change of the packet as that of the southwesterly jet located northwest of the packet. Therefore, when there is a southwesterly jet over the Tibet Plateau, the change in spatial scale will be much more obvious. From this discussion, we may conclude that when the Rossby wave packet passes over the plateau, its latitudinal scale will enlarge and the weather system will tend to develop more easily. This agrees with observational data (Zhu, Lin, and Shao, 1982).

Taking $a = b = 0.5$, we have computed the structural changes of the wave packet. The results, after integrating for 200 dimensionless time steps, are given in Figure 3.9a and the solid line C in Figure 3.8. It can be found from Figure 3.9a that, in such a case, the longitudinal scale will decrease about 10% and the latitudinal scale will increase 40%, after 200 dimensionless time steps. The trough line will tilt toward the Y-axis about 12° after 200 dimensionless time steps, as shown in Figure 3.8 by the solid line C.

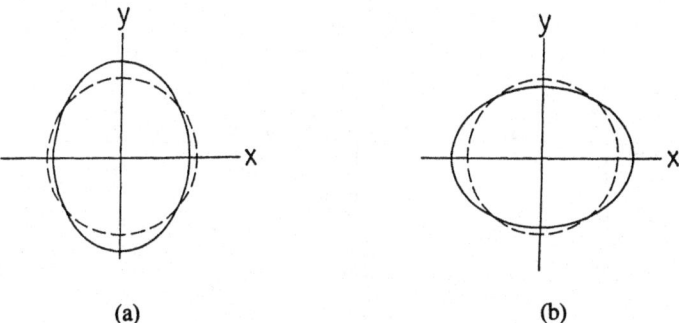

(a)   (b)

FIGURE 3.9. The spatial scale change of the Rossby wave packet on (a) convex and (b) concave topography. Key: - - -, an ideal streamline at $T = 0$; —, the ideal streamline at $T = 200$. After Yang (1987).

Concave Topography. This kind of topography can be more easily found in oceans than in the atmosphere. However, basins in both the ocean and the atmosphere may be considered to be this type of topography. We can discuss the effect of this type of topography using the same method used for the convex topography. Figure 3.9b and the solid line D in Figure 3.8 illustrate this effect upon the structural changes of the wave packet, where $\partial \beta_1 / \partial X = 0$, $\partial \beta_1 / \partial Y = 0.5$, $\partial \beta_2 / \partial X = 0.5$, and $\partial \beta_2 / \partial Y = 0$ have been taken. The computations show that after 200 dimensionless time steps, the wave packet's longitudinal scale has enlarged about 20%, while its latitudinal scale has shrunk about 10%, and the trough (or ridge) line will tilt westward about 7%.

In Figure 3.8 the dashed line is the trough (or ridge) line at the beginning, that is, at $T = 0$, and the solid lines A, B, C, and D are trough (or ridge) lines after 200 dimensionless time steps, corresponding to north–south, east–west, convex, and concave topographies, respectively.

3.6 Two Examples and a Simple Explanation

We now consider two examples of the entire evolution of the Rossby wave packet. In the first example, we look at a case in which the evolution of the Rossby wave packet takes place on the right side of a southwesterly jet over convex topography. According to our results, on the right side of a southwesterly jet over convex topography, the Rossby wave packet will evolve in such a manner that its longitudinal scale will become small and its latitudinal scale will become large, while the westward tilting trough (or ridge) line will tend to tilt toward the Y-axis. Figure 3.10 illustrates the entire evolution of the Rossby wave packet from model computations, when $\delta_0 = 0.2$, $\partial U / \partial Y = 0.02$, $\partial V / \partial X = -0.01$, $\partial \beta_2 / \partial X = -0.2$, and $\partial \beta_1 / \partial Y = -0.2$. We assume that the Rossby wave packet is westward tilting and that

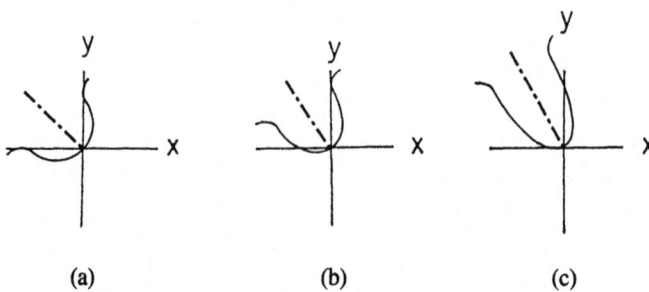

FIGURE 3.10. The evolution of the Rossby wave packet on the right side of a southwesterly jet and convex topography. (a) At $T = 0$, (b) at $T = 100$, and (c) at $T = 200$. Key: —, the ideal streamline; - · -, the trough line. After Yang (1987).

its streamline is always symmetric about its trough line. Initially, the wave packet is supposed to be located to the west of the topography, the trough line is tilted westward $45°$, and one of its streamlines has an ideal profile, as shown in Figure 3.10a, where the solid line is an ideal streamline and the heavy broken line is its trough line. After computing for 100 dimensionless time steps, that is, two and a half days of the dimensional time scale, Figure 3.10b is obtained. If we assume that, at this point, the wave packet has just arrived at the middle of the topography, then according to our arguments and computations in Section 3.3, the phase speed is accelerated during these 100 dimensionless time steps. It is found from the computations that the longitudinal scale has decreased about 10% and the latitudinal scale has increased about 40%. However, the whole spatial scale has increased by about 12% and the trough line has tilted $12°$ more toward the Y-axis. The pattern for the computation for another 100 dimensionless time steps, is shown in Figure 3.10c. During this period of time, the phase speed has been decreased. The longitudinal scale remains almost the same, while the latitudinal scale has increased dramatically, to almost two and a half times as large as it was at $T = 0$. Yet, the whole spatial scale of the wave packet has increased only about another 6% during these 100 dimensionless time steps, and the trough line has now been tilted a total of $13°$ toward the Y-axis, since $T = 0$.

Figure 3.11 is another example of the evolution of the Rossby wave packet when it is located on the left side of a southwesterly jet with the concave topography, and when $\delta_0 = 0.2$, $\partial U/\partial Y = -0.02$, $\partial V/\partial X = 0.01$, $\partial \beta_2/\partial X = 0.2$, and $\partial \beta_1/\partial Y = 0.2$. Figure 3.11b and Figure 3.11c show the results after computing for 100 dimensionless time steps and 200 dimensionless time steps from $T = 0$, respectively. From the figures, it can be seen that the longitudinal scale has increased 20% during the first 100 dimensionless time steps, and 50% for 200 dimensionless time steps, while the latitudinal scale has decreased about 10% for each of the 100 dimen-

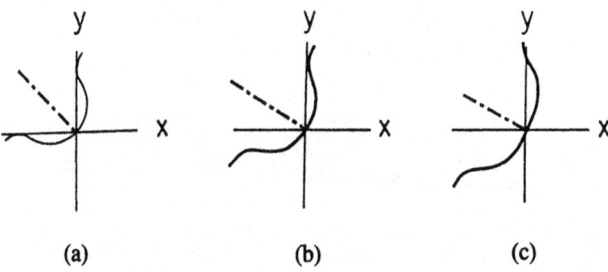

FIGURE 3.11. The same as Figure 3.10, except the packet evolves on the left side of a southwesterly jet and the topography is concave. After Yang (1987).

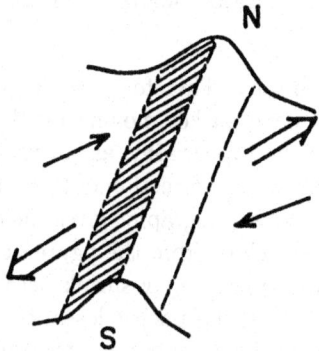

FIGURE 3.12. A simple interpretation of the effect of a north–south-oriented topography upon the evolution of the packet. The packet's latitudinal scale shrinks, since the energy propagates away from the packet. The solid arrow points in the wave vector direction; the double arrow is the direction of the energy propagation. After Yang (1987).

sionless time steps. However, the whole spatial scale of the wave packet has decreased only about 3% for each of the 100 time steps, while its trough line has further tilted 9° and 17° after 100 and 200 dimensionless time steps, respectively. Compared with the results in the first example, it is interesting to note that the speed of evolution in this case is different from that in the first case, though we used the same order of magnitude for each parameter in both cases.

Now, we try to demonstrate the mechanism of the change in the Rossby wave packet by the following two examples. First, suppose that there is a disturbance on a north–south-oriented type topography, as shown in Figure 3.12. From eqs. (3.45) to (3.48) in Section 3.3, it can be proven that on the west side of this topography, the phase velocity of the wave packet will be toward the northeast, whereas the energy will propagate toward the southwest. However, the phase velocity of the Rossby wave packet will be toward the southwest, while its energy will propagate toward the northeast

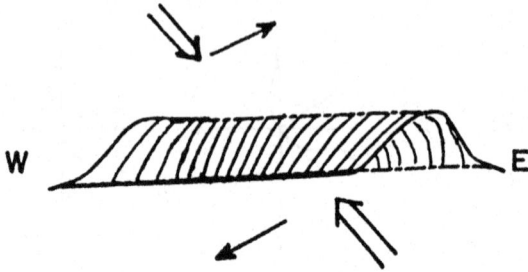

FIGURE 3.13. A simple interpretation of the effect of an east–west-oriented topography upon the evolution of the packet. The packet's longitudinal scale stretches, since the energy propagates toward the packet. After Yang (1987).

on the east side of topography. Therefore, in this case, the topography will result in the Rossby wave packet moving up the topography and its energy propagating down the topography, as shown in Figure 3.12, where the double arrow line points in the direction of the energy propagation and the solid arrow line, beside the topography, gives the direction of the phase velocity of the wave packet. Thus, from the figure, it can be seen that the latitudinal scale of the wave packet may shrink because the energy moves away from the packet, that is, the packet loses its energy into the basic flow, which is the only place the wave packet can transfer its energy in the present model. If the system is tilted westward, then the trough line will tend to the Y-axis because of the difference in phase velocity of the wave packet between the two sides of the topography.

Similarly, we can consider another case, namely, the evolution of the Rossby wave packet on an east–west-oriented topography, as shown in Figure 3.13. Since, in the present case, the energy moves toward the packet, that is, the wave packet gains energy from the basic flow, which is the only available energy source here, the wave packet's longitudinal scale will stretch, so that the whole spatial scale of the wave packet enlarges. The difference in the wave packet vectors between the two sides of the topography will result in the westward-tilting trough line tending toward the Y-axis; that is, the south–north orientation.

3.7   Conclusions

From the above discussions and some model computations, we draw the following main conclusions:

1. There are three constraints on the barotropic instability of the wave packet associated with three integral properties, which are analogues of the conservation of energy, wave-action, and enstrophy.

2. The wave packet's energy increases as the spatial scale (wave number) of the wave packet enlarges (decreases). The wave packet's energy decreases as the spatial scale (wave number) of the wave packet shrinks (increases).

3. The δ-effect results in a structural change in the packet. The Rossby wave packet's latitudinal scale increases and its westward tilting trough line tends toward the Y-axis, while the eastward tilting trough line tends toward the east. This is called the δ-effect. The computations show that the wole spatial scale has increased by about 8% and the westward tilting trough (or ridge) line has tilted 5° toward the Y-axis after 200 dimensionless time steps.

4. The Rossby wave packet, when it is located on the left (right) side of the southwesterly jet, will enlarge (shrink) its longitudinal scale and shrink (enlarge) its latitudinal scale. The westward (eastward) tilting trough line will tilt further westward (toward the Y-axis) when the packet is located on the left side of the jet. The westward (eastward) tilting trough line will tilt toward the Y-axis (the east) when located on the right side of the jet. The Rossby wave packet, when located on the left (right) side of a southeasterly jet, will enlarge (shrink) its longitudinal scale and latitudinal scale. The numerical computations show that the whole spatial scale of the wave packet could shrink (enlarge) by about 10% (12%) on the left side (right side) of a southwesterly jet after 200 dimensionless time steps. The westward tilting trough (or ridge) line could tilt up to 16° (18°) toward the west (the Y-axis) on the left side (right side) of a southwesterly jet. The whole spatial scale could enlarge (shrink) by 35% (25%) on the left side (right side) of a southeasterly jet after 200 dimensionless time steps. The westward tilting trough line could tilt about 8° (5°) toward the Y-axis (the west) on the left side (right side) of a southeasterly jet.

5. A linear topography will not alter the structure of the Rossby wave packet. However, the nonlinear topography does affect the structure of the packet. The results suggest that mountains, especially the Rocky Mountains, may decrease (increase) the longitudinal scale of the westward (eastward) tilting disturbance system. The north–south-oriented topography will only decrease the Rossby wave packet's longitudinal scale and the westward (eastward) tilting trough line will tilt toward the Y-axis (further eastward). The east–west-oriented topography will only increase the packet's latitudinal scale. The convex topography will shrink its longitudinal scale and enlarge its latitudinal scale. The east–west-oriented topography will result in the westward (eastward) tilting trough line tilting toward the Y-axis (further eastward). The concave topography will enlarge the Rossby wave packet's longitudinal scale and shrink its latitudinal scale, and the

westward (eastward) tilting trough line will tilt further westward (toward the Y-axis), as proven by the model computations. For instance, the computations show that the north–south (east–west) -oriented topography could result in the whole spatial scale shrinking (enlarging) by about 10% (8%) after 200 dimensionless time steps.

Appendix A. Proofs of Integral Properties of the Wave Packet

Equation (3.66) can be rewritten as follows:

$$\frac{\partial}{\partial T}\left(\frac{K^2}{2}|\Psi_0|^2\right) + \nabla \cdot \left(\frac{K^2}{2}|\Psi_0|^2 \mathbf{C}_g\right) = mn|\Psi_0|^2\frac{\partial U}{\partial Y}. \tag{3.115}$$

The first term on the left-hand side of eq. (3.115) represents the local change in wave packet energy. The second term on the left-hand side of the equation is the energy flux and the term on the right-hand side of the equation is the change in wave packet energy due to the basic flow, that is, the energy interaction between the wave packet and the basic flow.

Integrating (3.115) over the whole area (S) of the wave packet and assuming there is no net energy flux through the boundary of the whole area (S), we obtain

$$\frac{\partial}{\partial T}\int\int_{(S)}\left(\frac{K^2}{2}|\Psi_0|^2\right)dXdY = \int\int_{(S)} mn|\Psi_0|^2\frac{\partial U}{\partial Y}dXdY, \tag{3.116}$$

which is the first integral property of wave packet, that is, eq. (3.67).

In order to prove the second integral property, that is, eq. (3.68), we have to use the group velocity. In the present case, the group velocity (3.47) and (3.48) become

$$C_{gX} = U - \frac{B}{K^2} + \frac{2Bm^2}{K^4} \tag{3.117}$$

and

$$C_{gY} = \frac{2mnB}{K^4}. \tag{3.118}$$

The equation governing the total wave number, that is, eq. (3.61), can be written as

$$\frac{\partial K^2}{\partial T} + \mathbf{C}_g \cdot \nabla K^2 = -2mn\frac{\partial U}{\partial Y} + \frac{2mn}{K^2}\frac{\partial B}{\partial Y}, \tag{3.119}$$

since F is a constant parameter. Equation (3.66) can be rewritten as

$$\frac{\partial}{\partial T}\left(\frac{K^2}{2}|\Psi_0|^2\right) + \mathbf{C}_g \cdot \nabla\left(\frac{K^2}{2}|\Psi_0|^2\right) + \frac{K^2}{2}|\Psi_0|^2\nabla \cdot \mathbf{C}_g = mn|\Psi_0|^2\frac{\partial U}{\partial Y}. \tag{3.120}$$

Multiplying (3.119) by $K^2|\Psi_0|^2$ and (3.120) by $2K^2$, and adding them together, we obtain

$$K^2|\Psi_0|^2\frac{\partial K^2}{\partial T} + K^2\frac{\partial}{\partial T}(K^2|\Psi_0|^2) + K^2|\Psi_0|^2\mathbf{C}_g \cdot \nabla K^2$$

$$+ K^2\mathbf{C}_g \cdot \nabla(K^2|\Psi_0|^2) + K^4|\Psi_0|^2\nabla \cdot \mathbf{C}_g = 2mn|\Psi_0|^2\frac{\partial B}{\partial Y}.$$

It can be rewritten as

$$\frac{\partial}{\partial T}(K^4|\Psi_0|^2) + \nabla \cdot (K^4|\Psi_0|^2\mathbf{C}_g) - 2mn|\Psi_0|^2\frac{\partial B}{\partial Y} = 0. \qquad (3.121)$$

Dividing (3.121) by B, the equation reads

$$\frac{1}{B}\frac{\partial}{\partial T}(K^4|\Psi_0|^2) + \frac{1}{B}\nabla \cdot (K^4|\Psi_0|^2\mathbf{C}_g) - \frac{2mn}{B}|\Psi_0|^2\frac{\partial B}{\partial Y} = 0. \qquad (3.122)$$

Since B is only a function of Y, eq. (3.122) can be further written in the following form:

$$\frac{1}{B}\frac{\partial}{\partial T}(K^4|\Psi_0|^2) + \nabla \cdot \left(\frac{K^4|\Psi_0|^2}{B}\mathbf{C}_g\right) + \frac{K^4|\Psi_0|^2}{B}C_{gY}\frac{\partial B}{\partial Y}$$

$$- \frac{2mn}{B}|\Psi_0|^2\frac{\partial B}{\partial Y} = 0. \qquad (3.123)$$

By (3.118), (3.123) becomes

$$\frac{1}{B}\frac{\partial}{\partial T}(K^4|\Psi_0|^2) + \nabla \cdot \left(\frac{K^4|\Psi_0|^2}{B}\mathbf{C}_g\right) = 0. \qquad (3.124)$$

Again, integrating (3.124) over the whole area of the wave packet, we obtain

$$\iint_{(S)} \frac{1}{B}\frac{\partial}{\partial T}(K^4|\Psi_0|^2)dXdY = 0. \qquad (3.125)$$

This is eq. (3.68).

Multiplying (3.124) by $(U_r - U)$, where U_r is a constant reference of zonal flow, and adding it to eq. (3.115), we derive

$$\frac{\partial}{\partial T}\left(\frac{K^2}{2}|\Psi_0|^2\right) + \nabla \cdot \left(\frac{K^2}{2}|\Psi_0|^2\right) - mn\frac{\partial U}{\partial Y}|\Psi_0|^2$$

$$+ \frac{U_r - U}{2B}\frac{\partial}{\partial T}(K^4|\Psi_0|^2) + (U_r - U)\nabla \cdot \left(\frac{K^4|\Psi_0|^2}{2B}\mathbf{C}_g\right) = 0$$

or

$$\frac{\partial}{\partial T}(K^2|\Psi_0|^2) + \frac{U_r - U}{2B}\frac{\partial}{\partial T}(K^4|\Psi_0|^2) + \nabla \cdot \left\{\frac{(U_r - U)K^4}{2B}|\Psi_0|^2\mathbf{C}_g\right\}$$

$$+ \nabla \cdot \left(\frac{K^2}{2}|\Psi_0|^2\right) + \frac{K^4}{2B}|\Psi_0|^2\mathbf{C}_g \cdot \nabla U - mn|\Psi_0|^2\frac{\partial U}{\partial Y} = 0. \qquad (3.126)$$

Since U is the only function of Y, the last two terms can be written

$$\frac{K^4}{2B}|\Psi_0|^2 C_{gY} - mn|\Psi_0|^2 \frac{\partial U}{\partial Y}$$

and will be canceled by the use of (3.118). Therefore, (3.126) reads

$$\frac{\partial}{\partial T}\left(\frac{K^2}{2}|\Psi_0|^2\right) + \frac{U_r - U}{B}\frac{\partial}{\partial T}\left(\frac{K^4}{2}|\Psi_0|^2\right) + \nabla \cdot \left(\frac{K^2}{2}|\Psi_0|^2\right)$$
$$+ \nabla \cdot \left\{\frac{(U_r - U)}{2B}K^4|\Psi_0|^2 \mathbf{C}_g\right\} = 0. \qquad (3.127)$$

Integrating (3.127) over the whole area of wave packet gives the last integral property of the wave packet. This completes the proof.

References

Drazin, P.G., and Reid, W.H. (1981). *Hydrodynamic Stability.* Cambridge University Press, London.

Hayashi, Y., and Golder, D.G. (1983). Transient planetary waves simulated by GFDL spectral general circulation models. *J. Atmos. Sci.* **40**, 941–957.

Hoskins, B.J., and Karoly, D.J. (1981). The steady linear respose of a spherical atmosphere to thermal and orographic forcing. *J. Atmos. Sci.* **38**, 1179–1196.

Kuo, H.L. (1949). Dynamic instability of two-dimensional non-divergent flow in a barotropic atmosphere. *J. Meteorol.* **6**, 105–122.

Pedlosky, J. (1987). *Geophysical Fluid Dynamics,* 2nd ed. Springer-Verlag, New York.

Phillips, N.A. (1965). Elementary Rossby waves. *Tellus* **17**, 295–301.

Platzman, G.W. (1968). The Rossby wave. *Quart. J. Res. Meteorol. Soc.* **94**, 225–248.

Yang, H. (1987). Evolution of a Rossby wave packet in barotropic flows with asymmetric basic flow, topography and δ-effect. *J. Atmos. Sci.* **44**, 2267–2276.

Yang, H., and Yang, D. (1988). Jet stream and the stationary forcing Rossby wave packet in relation to the teleconnection in the atmosphere. *Acta Meteorol. Sinica* **46**, 403–411.

Yang, H., and Yang, D. (1990). Forced Rossby wave packet propagation and teleconnectins in the atmosphere. *Acta Meteorol. Sinica* **4**, 18–26.

Zeng, Q.-C. (1979). On nonlinear interaction of motion and process of rotational adaptation in rotating atmosphere. *Sci. Sinica* **22**, 945–957.

Zeng, Q.-C. (1982). On the evolution and interaction of disturbances and zonal flow in rotating barotropic atmosphere. *J. Meteorol. Soc. Japan* **60**, 24–31.

Zhu, Q., Lin, J., and Shao, S. (1982). *Synoptic Meteorology.* Chinese edition, Meteorology Press, Beijing.

4

Global Behavior: The Wave Packet Structural Vacillation

4.1 Introduction

In this chapter we further study the evolution of wave packet discussed in Chapter 3. In order to illustrate the global behavior of the evolution of the wave packet, we employ phase space and phase space diagrams. The phase space considered here is the space consisting of local wave numbers. This phase space has been called the *WKB phase space* (Yang, 1988a,b,c).

In the present case, however, it is the *WKB phase plane,* since there are only two local wave numbers, which consist of a plane in a phase space. The global behavior of the structural change in the wave packet due to the zonal, meridional, and asymmetric basic flow and the variety of the topography on the δ-surface of the earth are discussed and compared with those on the earth's β-plane, by the use of the WKB phase space. It shows that the governing system on the earth's δ-surface might be dynamically different from that on the earth's β-plane. We will further show that it is possible for the wave packet structural vacillation to exist on both the β-plane and the δ-surface. The wave packet structural vacillation is characterized by the time periodic change in the structure of the wave packet. Both the tilt and the spatial scales of the wave packet simultaneously vary periodically with time. The wave packet's structural vacillation is also characterized by closed trajectories on the WKB phase plane. The wave packet's structural vacillation suggests a possible mechanism of vacillations observed in geophysical fluids (Yang, 1988a).

In this chapter, we still use the same model as we used in Chapter 3. The global behavior in the presence of various basic flows is addressed in Section 4.2 on the β-plane. The results on the δ-surface are presented in Section 4.3. Section 4.4 is devoted to the topography and the final section is a summary of the results.

4.2   Behavior due to the Basic Flows on the Earth's β-Plane

When only the effect of the main part of the basic flow is considered on the earth's β-plane, the governing equations (3.50) to (3.51) will become as follows:

$$\frac{D_g m}{DT} = -\frac{n}{K^2}(m^2 + n^2)\frac{\partial V}{\partial X}, \tag{4.1}$$

$$\frac{D_g n}{DT} = -\frac{m}{K^2}(m^2 + n^2)\frac{\partial U}{\partial Y}, \tag{4.2}$$

$$\frac{D_g}{DT}(m^2 + n^2) = -2mn\frac{m^2 + n^2}{K^2}\left(\frac{\partial U}{\partial Y} + \frac{\partial V}{\partial X}\right), \tag{4.3}$$

and

$$\frac{D_g}{DT}\left(-\frac{n}{m}\right) = \left(\frac{\partial U}{\partial Y} - \frac{n^2}{m^2}\frac{\partial V}{\partial X}\right)\frac{m^2 + n^2}{K^2}. \tag{4.4}$$

4.2.1   ZONAL BASIC FLOW

On the zonal basic flow, it can be shown from (4.1) to (4.4) that the Rossby wave packet's longitudinal scale will not change, though the westward tilting wave packet's latitudinal scale increases (decreases) when the wave packet is located on the right side (left side) of the jet center. Since the energy of the Rossby wave packet always propagates with its group velocity, we can take the group velocity as the characteristic direction and then integrate the wave packet along that characteristic line. It can be readily proven that there are simple wave packet solutions, which are related by

$$m = m_0 = \text{constant} \tag{4.5}$$

and

$$n + \frac{F}{m}\arctan\frac{n}{m} = -m\frac{\partial U}{\partial Y}T + C, \tag{4.6}$$

where C is the integration constant to be determined by the initial conditions.

4.2.2   MERIDIONAL BASIC FLOW

Meridional basic flow may easily be observed in the ocean, especially near the north–south-oriented coasts. From (4.1) to (4.4), it can be shown that the meridional basic flow increases (decreases) the westward tilting Rossby wave packet's longitudinal scale when it is located on the right side (left side) of the jet center. The meridional basic flow does not alter the packet's latitudinal scale. Integrating along the characteristic line, we have

$$m + \frac{F}{n} \arctan \frac{m}{n} = -n\frac{\partial V}{\partial X}T + C, \tag{4.7}$$

where

$$n = n_0 = \text{constant} \tag{4.8}$$

and C is the integration constant to be determined by the initial condition.

The results show that, on the earth's β-plane, the property of the dynamic effect of the meridional basic flow is similar to that of the zonal basic flow. The only difference is that the zonal basic flow changes the packet's latitudinal scale, whereas the meridional basic flow alters the longitudinal scale.

4.2.3   ASYMMETRIC BASIC FLOW

From (4.1) to (4.2), integrating along the characteristic line of the wave packet, we can obtain the WKB trajectories of the Rossby wave packet for the supposed basic flow as follows:

$$\frac{\partial U}{\partial Y}m^2 - \frac{\partial V}{\partial X}n^2 = C. \tag{4.9}$$

The wave packet solutions can also be obtained, as shown in Appendix A.

Figure 4.1 shows the global behavior of the evolution of a Rossby wave packet by the WKB trajectories on the WKB phase plane in the asymmetric basic flow in which $(\partial U/\partial Y)(\partial V/\partial X) \leq 0$. The southwesterly jet, which often occurs above southeast Asia in the atmosphere, is a typical example of this case. Figure 4.1a is the case in which $(\partial U/\partial Y) < 0$, where the two straight lines correspond to the case of the zonal basic flow. Figure 4.1b is the case in which $(\partial V/\partial X) < 0$, where the two straight lines correspond to the case of the meridional basic flow. The arrows in the WKB phase plane indicate the flow direction of the evolution of the wave packet. Figure 4.2a shows the evolution of the whole scale of the packet, which is defined by $2\pi/(m^2 + n^2)^{\frac{1}{2}}$. Figure 4.2b and c illustrate the evolution of the packet's local wave numbers along the X- and Y-direction, respectively. Figure 4.3 is the evolution of the tilt of the wave packet. In Figure 4.2 and Figure 4.3, the wave packet is set to begin with $m = n = 6$ and parameters taking $\partial U/\partial Y = -0.02$, $\partial V/\partial X = 0.01$, and $F = 1.0$. That $\partial U/\partial Y = -0.02$ and $\partial V/\partial X = 0.01$ corresponds to the basic flow changes of about 2 m/s in the U component and 1 m/s in the V component, 100 km out from the jet center. Here $F = 1.0$ corresponds to the case in which the wave length of the disturbance system in the atmosphere is about 3000 to 4000 km in the middle latitude regions. The figures show that the whole scale and the tilt of the Rossby wave packet will change periodically with time. We call this phenomenon the *wave packet's structural vacillation*. From the

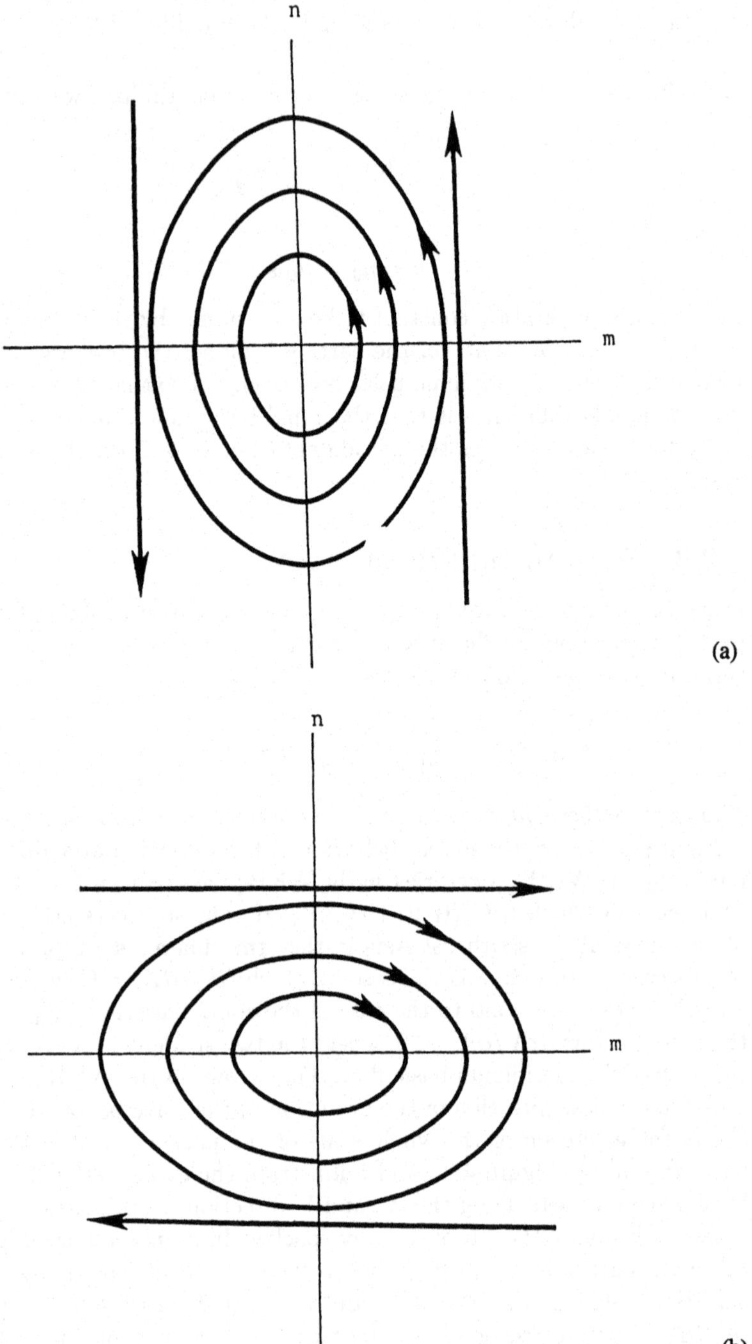

(a)

(b)

FIGURE 4.1. The WKB trajectories on the WKB phase plane on the earth's β-plane in the presence of basic flows, where $(\partial U/\partial Y)(\partial V/\partial X) \leq 0$. (a) $(\partial U/\partial Y) < 0$ and (b) $(\partial U/\partial Y) > 0$. The arrows indicate the flow direction of the evolution of the wave packet. Two straight lines correspond to the case in the zonal basic flow (a) and in the meridional basic flow (b). The ordinate is the Y-directional local wave number, and the abscissa is the local wave number in the X-direction. After Yang (1988a).

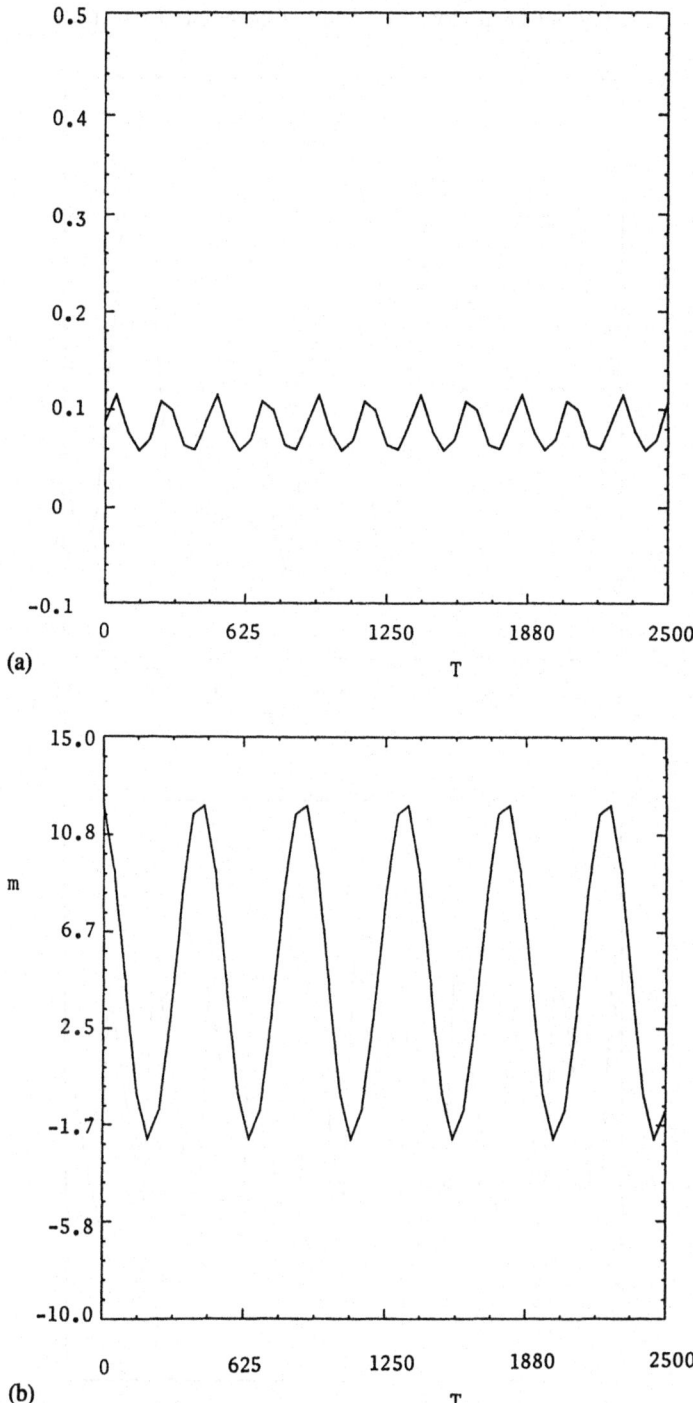

(a)

(b)

FIGURE 4.2. The evolution of a Rossby wave packet on the earth's β-plane in the case $(\partial U/\partial Y) = -0.02$, $(\partial V/\partial X) = 0.01$, and $F = 1.0$. (a) The whole spatial scale, where the ordinate is the whole spatial scale; (b) the local wave number along the X-direction; and (c) the local wave number along the Y-direction. The abscissa is the time. After Yang (1988a).

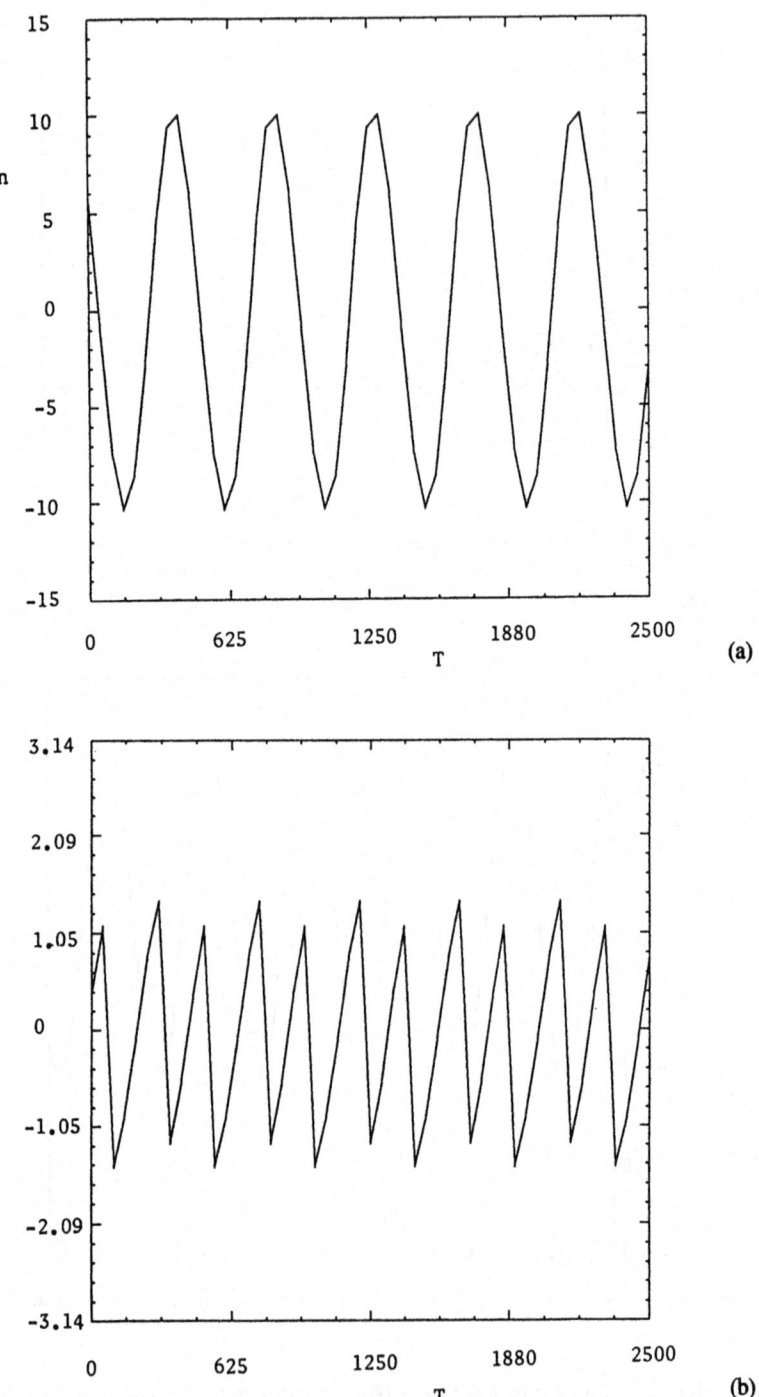

FIGURE 4.3. The evolution of the tilt of the Rossby wave packet, corresponding to Figure 4.2, where the ordinate is the tilt and the abscissa is the time. After Yang (1988a).

figures, we find that the wave packet's structural vacillation is characterized by the simultaneous periodic changes in the tilt, the scale along the X-direction, the scale along the Y-direction, and the whole scale of the wave packet. In the case of Figure 4.2 and Figure 4.3, it can be found that the period of the wave packet's structural vacillation is about 450 dimensionless time steps, which is about 11 to 12 days in the dimensional time scale, a typical order-of-magnitude of time scale for weather systems. Figure 4.4 is an example of the evolution of a Rossby wave packet in a southwesterly jet on the earth's β-plane, starting with $m = m = 6$ and parameters taking $\partial U/\partial Y = 0.02$, $\partial V/\partial X = -0.01$, and $F = 1$. In Figure 4.4, the solid line and the dashed line are an ideal streamline and the trough line of the wave packet, respectively. From these figures, it can be seen that the packet first tilts north–west to south–east, $m = n = 6$, at $T = 0$ (Figure 4.4a) and then north to south, $m = 0$, $n = 10$, reaching its maximum longitudinal scale at

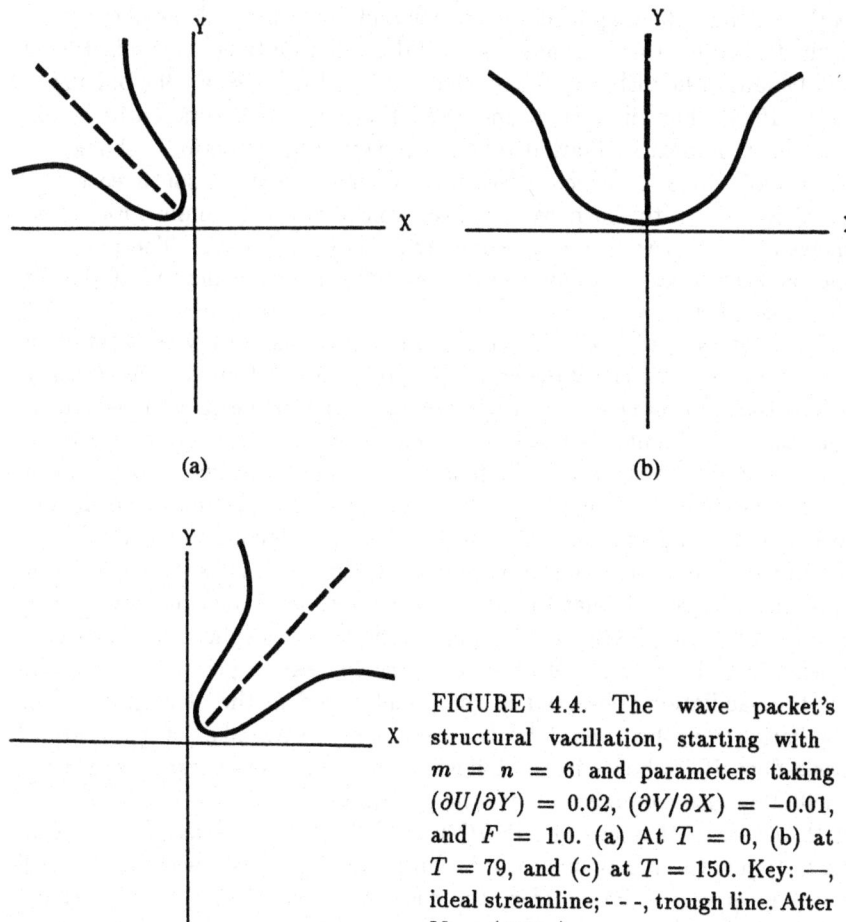

(a)

(b)

(c)

FIGURE 4.4. The wave packet's structural vacillation, starting with $m = n = 6$ and parameters taking $(\partial U/\partial Y) = 0.02$, $(\partial V/\partial X) = -0.01$, and $F = 1.0$. (a) At $T = 0$, (b) at $T = 79$, and (c) at $T = 150$. Key: —, ideal streamline; - - -, trough line. After Yang (1988a).

$T = 79$ (Figure 4.4b); it later tilts north–east to south–west, $m = -6$, $n = 6$, at $T = 150$ (Figure 4.4c).

This phenomenon is reminiscent of the vacillations in the annulus and the atmosphere. When daily weather maps are examined, one can readily find that some weather patterns will repeat in a quasi-periodic manner. The index cycle is a typical example; it has about a 13-day period. It is well known that the steady basic flow in the atmosphere is mainly driven by topography and heating. This basic flow is not zonally symmetric. From this observation, we know that the basic flow in the atmosphere is a wave-like jet at the climate average position. In addition, in a study of 9.5 years of the daily NMC (National Meteorological Centre) gridded height and temperature field, McGuirk and Reiter (1976) found a strong, persistent, significant oscillation of about a 24-day periodicity in the energy parameter during the winter season. Gruber (1975) has observed a 14- to 16-day periodicity in the tropics. A 18- to 23-day period vacillation was also found by Webster and Keller (1975) in the EOLE data from the Southern Hemisphere. However, in the annulus, the amplitude vacillation and the structural vacillation have been studied for three decades (e.g., Hide, 1958; Fultz et al., 1959; Pfeffer and Chiang, 1967; Elberry, 1968; Pfeffer et al., 1974, 1980a, 1980b; Buzyna et al., 1984). The tilt trough line vacillation may be considered to be one form of structural vacillation that is characterized by periodic changes in the tilt of waves, in which waves first tilt north–west to south–east, and later, north–east to south–west, relative to the rotating annulus. Figure 4.5 shows a series of results from the annulus (Elberry, 1968). The period of the vacillation here is about 16 rotations, which is equivalent to 16 days in the atmosphere.

By comparing Figure 4.4 with Figure 4.5, one can easily see that there is a qualitative similarity between the two figures. Although, theoretically, a flow that is averaged over a long enough time in the annulus is zonally symmetric; in reality, however, the time-mean flow will never be zonally symmetric due to the aperiodic motions and the limitation of observation time in annulus experiments. If the basic flow is considered to be the one averaged over a period longer than the period of vacillation, then it will be found that the basic flow is a wake-like jet, that is, it is asymmetric. From this analysis, we find that the orders of magnitude of the time period scale and the wave length scale in the annulus and the atmosphere agree with our predictions. Therefore, the present theory suggests a possible mechanism for the vacillations observed in the annulus and in the atmosphere. The vacillation may be caused by the asymmetric wave-like jet structure of basic flow. It is shown in the following that the wave packet's structural vacillation also can exist under other conditions.

However, in the case of $(\partial Y/\partial Y)(\partial V/\partial X) > 0$, for example, a southeasterly jet, the behavior of the evolution of the Rossby wave packet will be qualitatively different from the case of the southwesterly jet. Figure 4.6 illustrates the behavior of the packet in such a case on the WKB phase

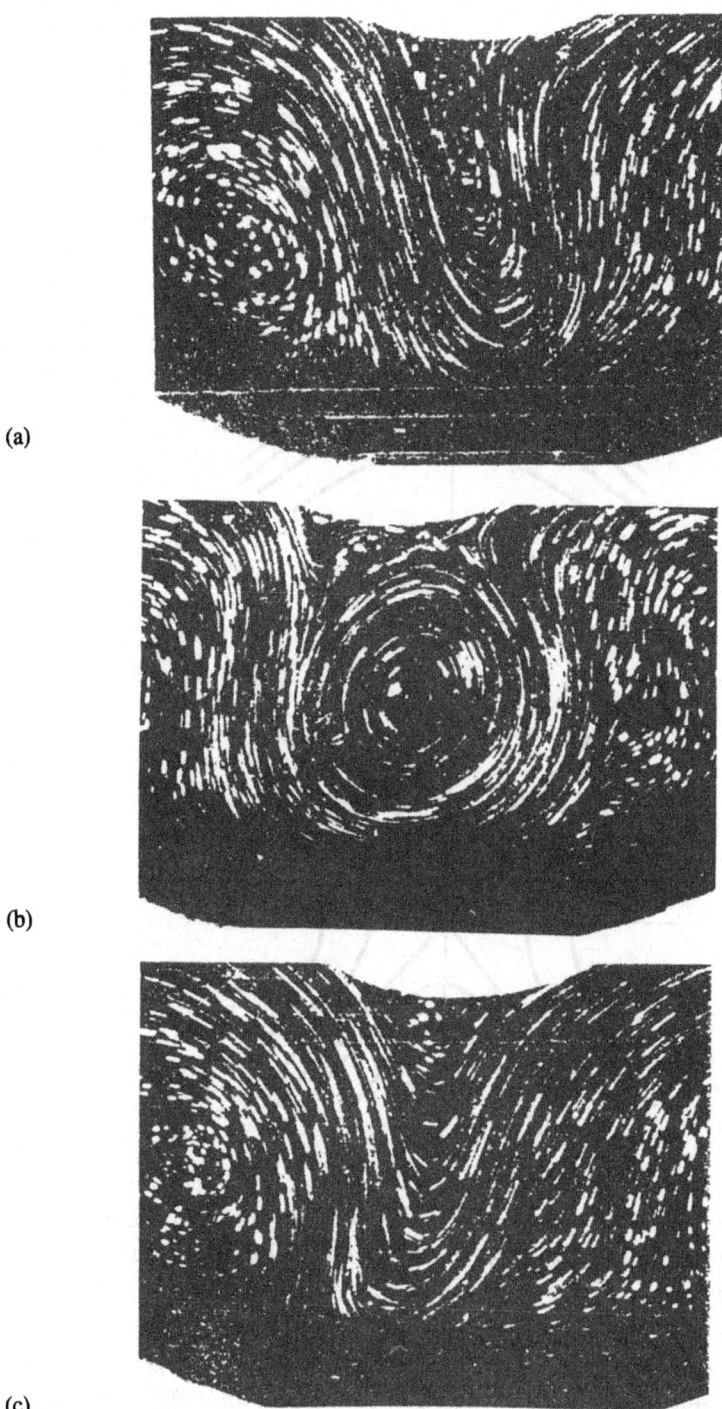

(a)

(b)

(c)

FIGURE 4.5. The annulus result of wave vacillation from Elberry (1968). Photographs of top surface aluminum powder streaks at alternate rotations illustrate the wave vacillation with a period of 16 rotations. (a) $T = 1$, (b) $T = 5$, and (c) $T = 15$.

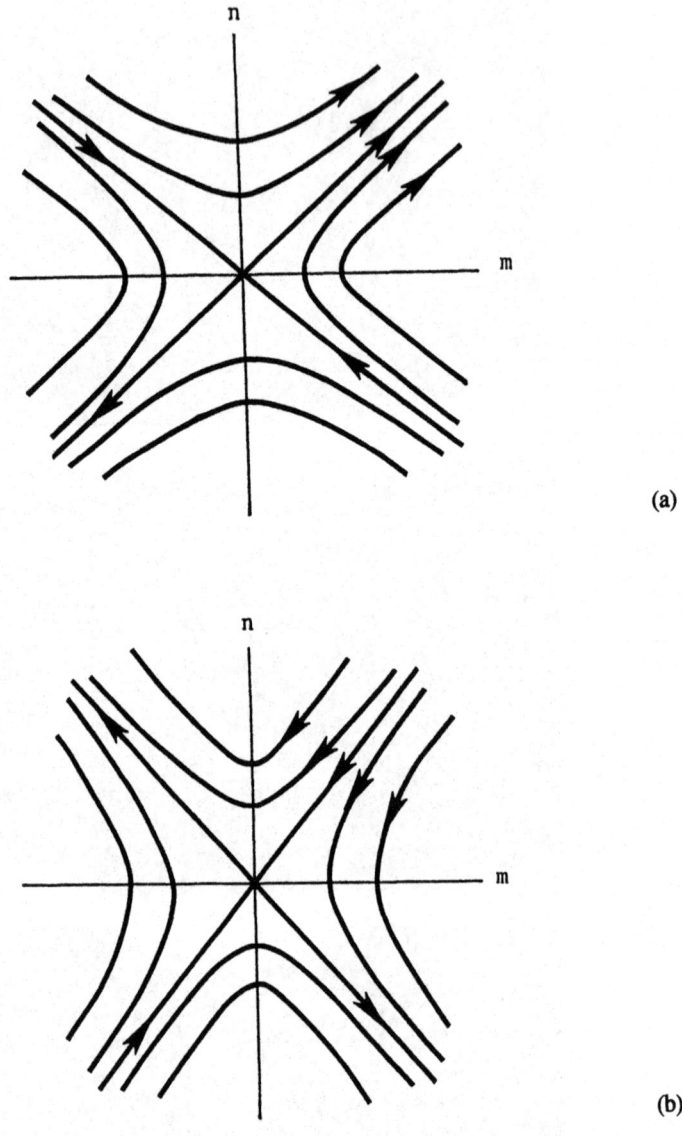

FIGURE 4.6. The WKB phase plane on the earth's β-plane in the case of a southeasterly jet. (a) On the right side and (b) on the left side of the jet. After Yang (1988a).

plane. Figure 4.6a shows the results when the Rossby wave packet is located on the right side of a southeasterly jet, in the case of $|\partial U/\partial Y| \leq |\partial V/\partial X|$ and $(\partial Y/\partial Y) < 0$. Figure 4.6b corresponds to the results when the packet is located on the left side of a southeasterly jet, in the case of $|\partial U/\partial Y| \geq |\partial V/\partial X|$ and $(\partial U/\partial Y) > 0$. Again, the arrows in the figures indicate the flow direction of the evolution of the wave packet on the WKB phase plane.

From the WKB phase plane, we can clearly see the global behavior of the evolution of a Rossby wave packet. The whole scale and the tilt of the packet evolve along the WKB trajectory in the arrow direction on the WKB phase plane with time. Comparing these results of Figure 4.6 with those of Figure 4.1, we can readily see that there is a qualitative difference between the effect of a southwesterly jet and that of a southeasterly jet. In a southwesterly jet, there is the wave packet's structural vacillation; whereas in a southeasterly jet, since the longitudinal scale and latitudinal scale increases (or decreases) simultaneously, there is no such structural vacillation. Therefore, the two systems are dynamically different, since on the WKB phase plane one system is characterized by a open line, while another system is characterized by a closed line.

4.3 Behavior due to the Basic Flow on the Earth's δ-Surface

It can be easily shown that the β-effect only affects the propagation properties of the Rossby wave packet and will not affect its structure. However, the δ-effect, which refers to the effect of the second derivative of the Coriolis parameter with respect to latitude, not only affects the propagation properties of the packet, it also alters the structure of a Rossby wave packet. This effect may be of great importance near the earth's poles, where $\beta = 0$. It had been shown in Chapter 3 that the δ-effect causes the Rossby wave packet's latitudinal scale to increase and its westward tilting troughline to tend toward the Y-axis.

If we only consider the δ-effect in (3.49) and (3.50), integrating along the characteristic line, we can readily obtain the wave packet solutions that are related by the following relation:

$$m = m_0 = \text{constant} \tag{4.10}$$

and

$$(F + m^2)n + \frac{1}{3}n^3 = -\delta_0 mT + C, \tag{4.11}$$

where C is the constant of integration to be determined by the initial condition. Equations (4.10) and (4.11) describe all the possibilities of the δ-effect upon the structural changes of a Rossby wave packet.

Because of the special property of the δ-effect, it is natural to consider the dynamic behavior of a Rossby wave packet on the earth's δ-surface. In what follows, we discuss the global behavior of the evolution of the packet on the earth's δ-surface and compare those on the earth's β-plane in the presence of a variety of basic flows. The equations governing the evolution of a Rossby wave packet on the δ-surface in the presence of the basic flow can be obtained from (3.50) to (3.53), that is,

$$\frac{D_g m}{DT} = -n \frac{m^2 + n^2}{K^2} \frac{\partial V}{\partial X}, \tag{4.12}$$

$$\frac{D_g n}{DT} = -\frac{m}{K^2} \left\{ \delta_0 + (m^2 + n^2) \frac{\partial U}{\partial Y} \right\}, \tag{4.13}$$

$$\frac{D_g}{DT}(m^2 + n^2) = -\frac{2mn}{K^2} \left\{ \delta_0 + (m^2 + n^2) \left(\frac{\partial U}{\partial Y} + \frac{\partial V}{\partial X} \right) \right\}, \tag{4.14}$$

and

$$\frac{D_g}{DT} \left(-\frac{n}{m} \right) = \frac{1}{K^2} \left\{ \delta_0 + \left(\frac{\partial U}{\partial Y} - \frac{n^2}{m^2} \frac{\partial V}{\partial X} \right) (m^2 + n^2) \right\}. \tag{4.15}$$

Equations (4.12) to (4.15) will be reduced to those obtained by Zeng (1982) when only the zonal basic flow is present. The asymmetric basic flow has been considered by Karoly (1983). He obtained a complete set of equations for the zonally varying basic flow on a spherical geometry in a barotropic model, which is similar to the present one. We discuss his results in Chapter 8. In his results, if we consider that the basic state is geostrophic and that the δ-surface approximation substitutes for the spherical geometry, then the present results can be derived from his, provided only the main part of basic flow is included. A comparison of the two results reveals the advantage of the δ-surface approximation. Under such an approximation, the system of equations becomes simply an ordinary system with constant coefficients for the supposed basic flow. However, in his results, this is impossible, due to the variation of second derivative of mean potential vorticity in latitude even for a supposed basic flow.

4.3.1   ZONAL BASIC FLOW

In the zonal flow, (4.12) shows that the Rossby wave packet's longitudinal scale does not change, though the packet's latitudinal scale does. Again, integrating along the characteristic line, we have the following wave packet solutions:

$$m = m_0 = \text{constant} \tag{4.16}$$

and

$$n + \frac{F - \delta_0 / \frac{\partial U}{\partial Y}}{\left(\delta_0 / \frac{\partial U}{\partial Y} + m^2 \right)^{\frac{1}{2}}} \arctan \frac{n}{\left(\delta_0 / \frac{\partial U}{\partial Y} + m^2 \right)^{\frac{1}{2}}} = -m \frac{\partial U}{\partial Y} T + C, \tag{4.17}$$

when

$$\frac{\partial U}{\partial Y} > -\frac{\delta_0}{m^2}; $$

or

$$m = m_0 = \text{constant} \tag{4.18}$$

and

$$n + \frac{F - \delta_0/\frac{\partial U}{\partial Y}}{2\left|\delta_0/\frac{\partial U}{\partial Y} + m^2\right|^{\frac{1}{2}}} \ln \frac{n + \left|m^2 + \delta_0/\frac{\partial U}{\partial Y}\right|^{\frac{1}{2}}}{n - \left|m^2 + \delta_0/\frac{\partial U}{\partial Y}\right|^{\frac{1}{2}}} = m - \frac{\partial U}{\partial Y}T + C, \quad (4.19)$$

when

$$\frac{\partial U}{\partial Y} < -\frac{\delta_0}{m^2}.$$

By comparing these results with those from the earth's β-plane, we can show that when $(\partial U/\partial U) > -(\delta_0/m^2)$, the qualitative properties of the evolution of a Rossby Wave packet on the δ-surface is similar to that on the β-plane. However, when $(\partial U/\partial Y) < -(\delta_0/m^2)$, the qualitative properties on the δ-surface will differ from those on the β-plane.

4.3.2 MERIDIONAL BASIC FLOW

In the meridional basic flow, the equations governing the evolution of a packet on the earth's δ-surface will become

$$\frac{D_g m}{DT} = -n\frac{m^2 + n^2}{K^2}\frac{\partial V}{\partial X}, \quad (4.20)$$

$$\frac{D_g n}{DT} = -\frac{m}{K^2}\delta_0, \quad (4.21)$$

$$\frac{D_g}{DT}(m^2 + n^2) = -\frac{2mn}{K^2}\left\{\delta_0 + (m^2 + n^2)\frac{\partial V}{\partial X}\right\}, \quad (4.22)$$

and

$$\frac{D_g}{DT}\left(-\frac{n}{m}\right) = \frac{1}{K^2}\left\{\delta_0 - \frac{n^2}{m^2}(m^2 + n^2)\frac{\partial V}{\partial X}\right\}. \quad (4.23)$$

From these equations, we can find that, on the δ-surface, the effect of a meridional basic flow upon the evolution of a packet will be qualitatively different from that on the β-plane. On the δ-surface, the longitudinal scale and the latitudinal scale of the packet change simultaneously. But on the β-plane, the latitudinal scale of the packet does not change with time. The dynamic effect of the meridional basic flow on the packet is also qualitatively different from that of the zonal basic flow on the δ-surface, which is different on the β-plane.

Figure 4.7 shows the global behavior of the packet in the meridional basic flow on the δ-surface. The results demonstrate that, in this case, the governing system on the earth's δ-surface is dynamically different from that on the earth's β-plane. In Figure 4.7, the solid lines are the WKB trajectories, whereas the two straight lines paralleling the m-axis are the WKB trajectories on the β-plane. Figure 4.7a corresponds to the case on the right side of the meridional basic flow, that is, $(\partial V/\partial X) < 0$, whereas

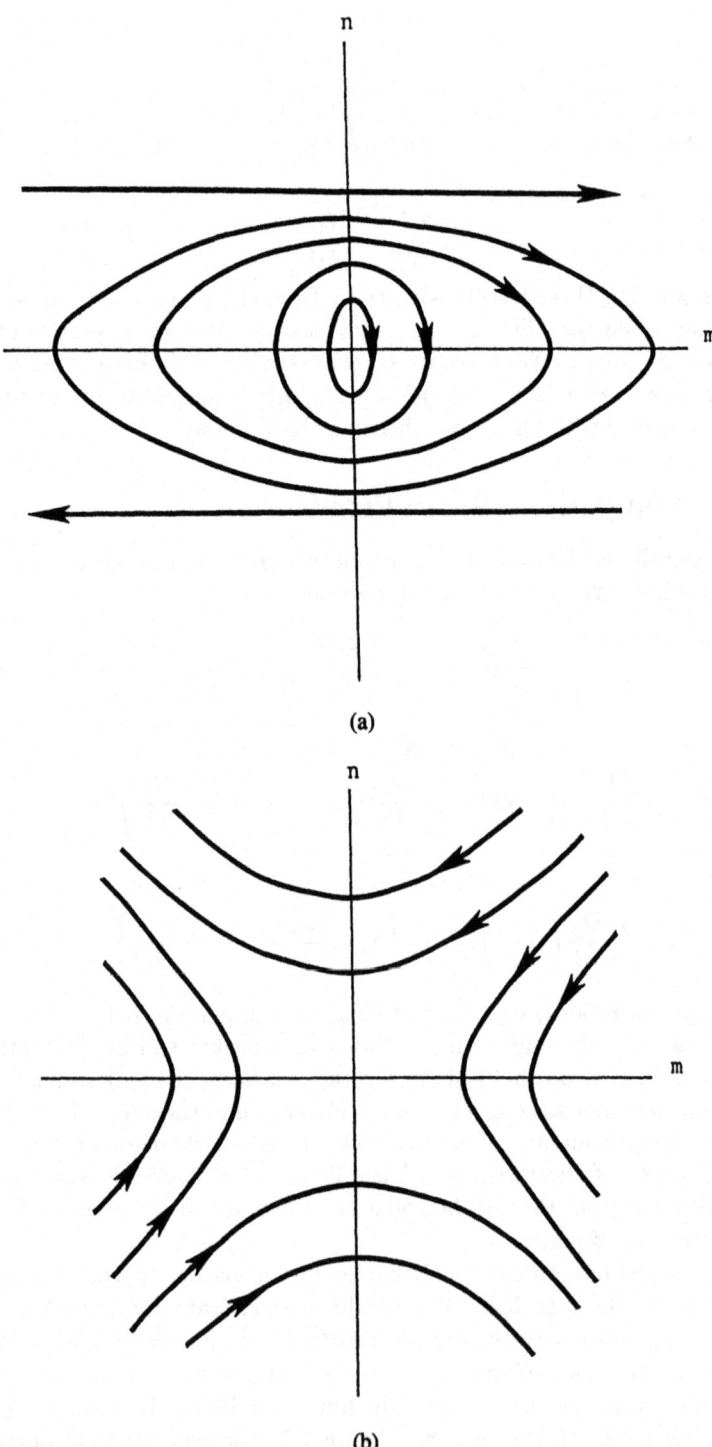

(a)

(b)

Figure 4.7b corresponds to the case on the left side of the meridional basic flow, that is, $(\partial V/\partial X) > 0$. On the δ-surface, the effect of the meridional basic flow on the left side is essentially different from that on the right side of the meridional basic flow.

On the right side, the WKB trajectory of the wave packet on the WKB phase plane is a closed line. Therefore, the evolution of the packet varies periodically with time. The results suggest that on the right side of the meridional basic flow, the evolution of a Rossby wave packet exhibits the wave packet's structural vacillation on the δ-surface of the earth, whereas it is impossible on the earth's β-plane. Figure 4.8 shows the results of such a wave packet structural vacillation in the case $(\partial V/\partial X) = -0.01$, $\delta_0 = 0.2$, and $F = 1.0$, starting from $m = n = 6$. Here, $(\partial V/\partial X) = -0.01$ corresponds to the meridional basic flow change of about $1\,\mathrm{m/s}$, $100\,\mathrm{km}$ out from the jet center; $\delta_0 = 0.2$ and $F = 1.0$ correspond to the case in which the wave length of the disturbance system is about 3000 to $4000\,\mathrm{km}$ in the middle latitude regions of the atmosphere. Figure 4.8a is the WKB trajectory of the evolution on the WKB phase plane. Figure 4.8b is the whole spatial scale of the packet changing with time. Figure 4.8c shows the time change of the tilt of the packet. Figures 4.8d and 4.8e correspond to the time changes of the packet's local wave numbers in the X-direction and Y-direction, respectively. We could find from the results in this case that the period of the packet's structural vacillation is larger than that in the case of the asymmetric basic flow on the β-plane under the present conditions.

The WKB solutions for this wave packet can be found in the WKB phase space as

$$m^2 + n^2 = \frac{1}{a} + Ce^{-an^2}, \tag{4.24}$$

where

$$a = \frac{\partial V}{\partial X}/\delta_0 \tag{4.25}$$

and C is a constant of integration to be determined by the initial condition of the wave packet. Thus, there are closed orbits in the WKB phase space, as is easily shown in the polar coordinates, that is,

$$r^2 = \frac{1}{a} + Ce^{-ar^2\cos^2\theta}. \tag{4.26}$$

The case $C = 0$ corresponds to a circle in the WKB phase space.

FIGURE 4.7. The WKB phase plane on the earth's δ-surface in the meridional basic flow. (a) On the right side and (b) on the left side of the jet. After Yang (1988a).

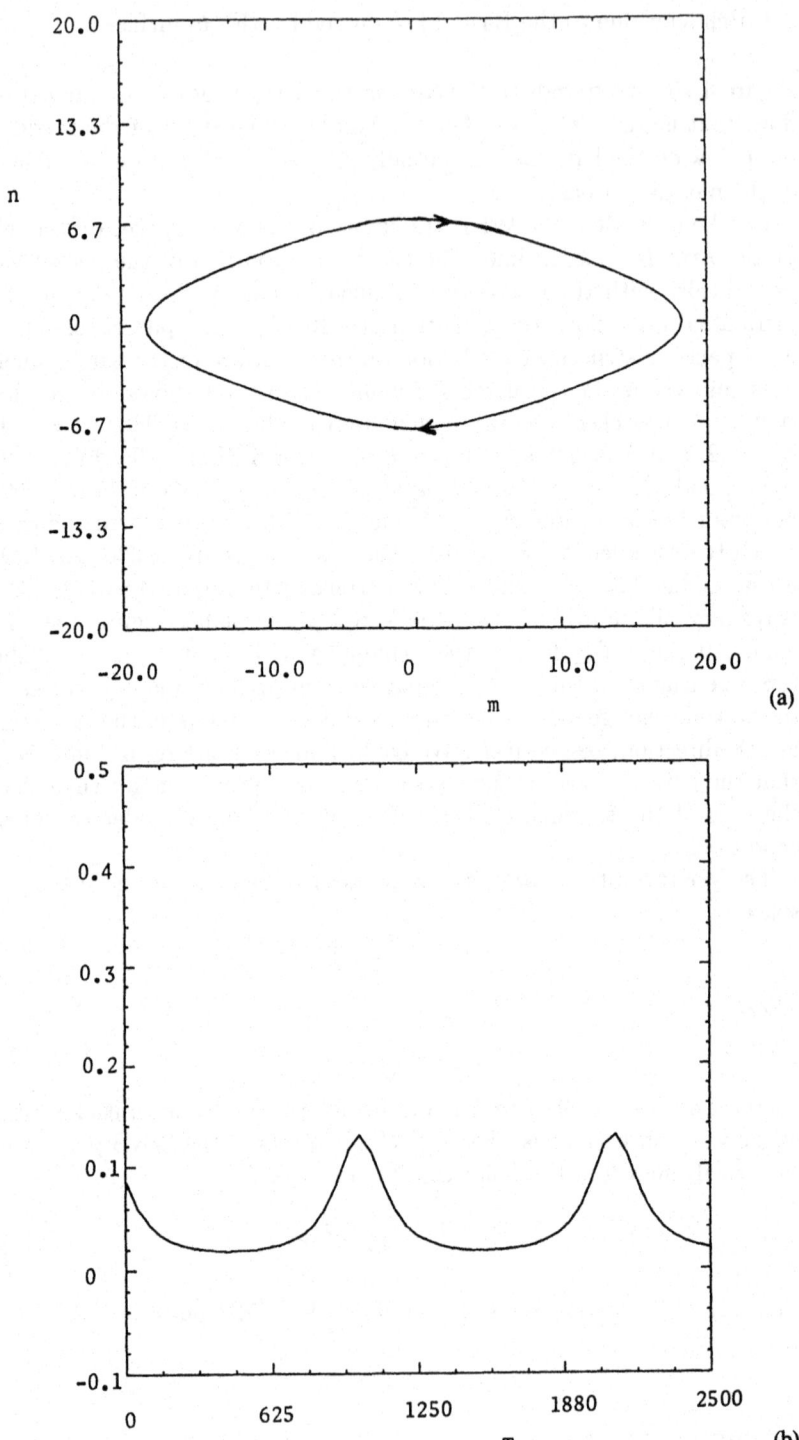

FIGURE 4.8. The wave packet's vacillation on the earth's δ-surface in the presence of a meridional basic flow. (a) The WKB phase plane, (b) the time change of the whole spatial scale,

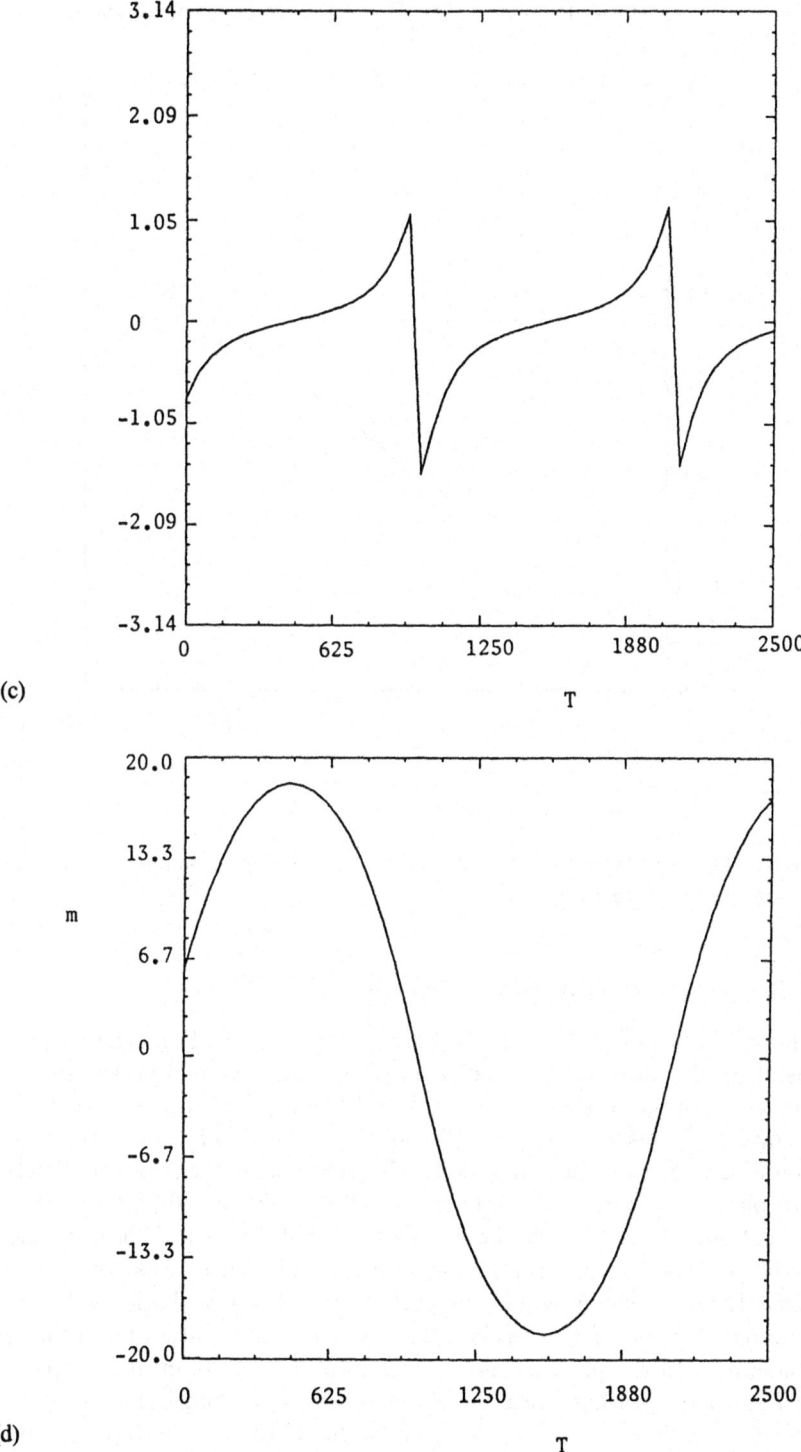

FIGURE 4.8 (cont.) (c) the time change of the tilt, (d) the time change of the local wave number along the X-direction, and

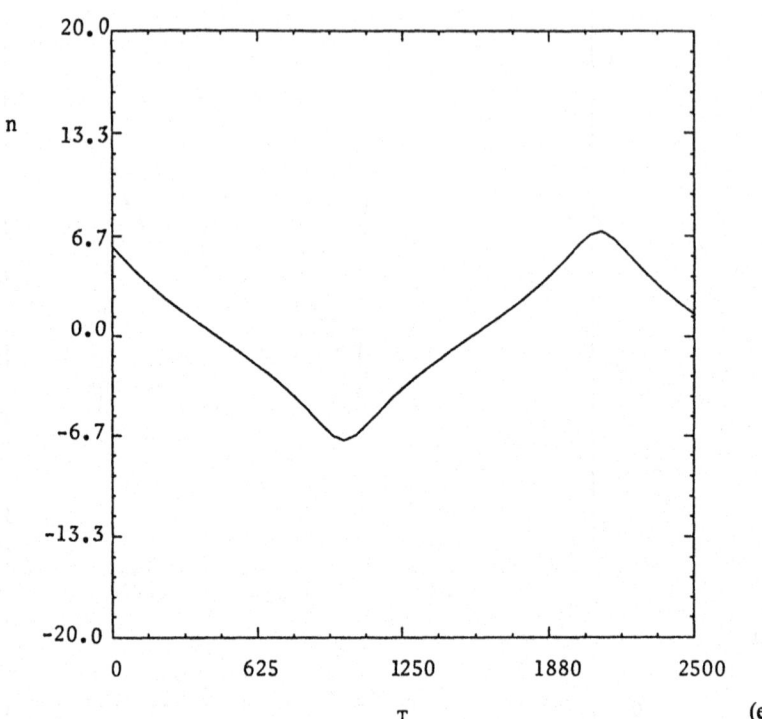

FIGURE 4.8 (cont.) (e) the time change of the local wave number along the Y-direction. After Yang (1988a).

4.3.3 ASYMMETRIC BASIC FLOW

In this case, the governing equations are (4.12) to (4.15). From these equations, it can be found that the behavior of the evolution of a packet on the δ-surface differs from that on the β-plane because of the presence of δ_0 in the governing equations. Figure 4.9 shows the global behavior of the evolution of a packet on the δ-surface by the WKB trajectories on the WKB phase plane, and parameters taking $(\partial U/\partial Y) = -0.02$, $(\partial V/\partial X) = 0.01$, $\delta_0 = 0.2$, and $F = 1.0$. The figure suggests that, in such a case, it may be still possible for the packet's structural vacillation to exist on the δ-surface. However, the period of the packet's structural vacillation will now be altered because of the δ-effect. The δ-effect complicates the problem considerably. In the present case, we are also able to obtain its complete integrals by integrating along the characteristic line. And from the equations, one can show that there are still periodic solutions once the condition $(\partial U/\partial Y)(\partial V/\partial X) < 0$ is satisfied.

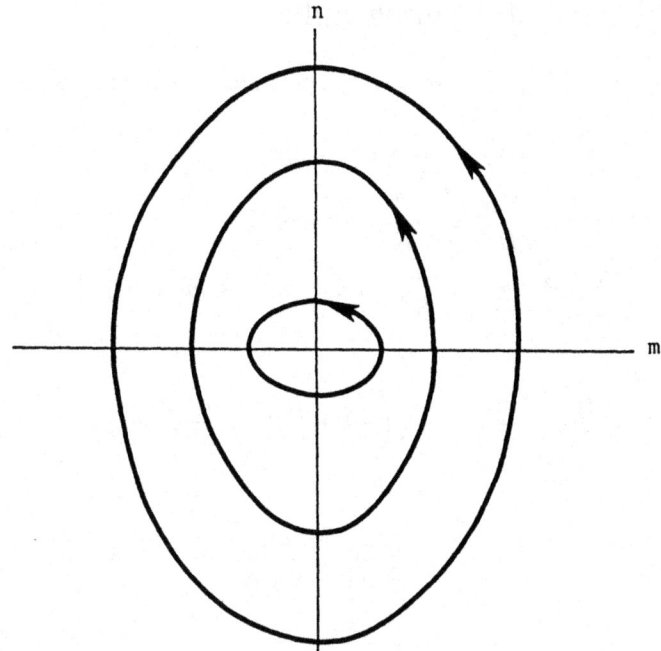

FIGURE 4.9. The WKB phase plane on the earth's δ-surface in the presence of a southwesterly basic flow. After Yang (1988a).

The complete WKB solutions are found to be

$$[k_1^2(m^2 + n^2)^2 - \delta_0^2] = -2k_1\delta_0 m^2 + C_1, \tag{4.27}$$

when $k_1 = k_2 = k$; and

$$k_2(m^2 + n^2) + \delta_0 + \frac{k_2}{k_1 - k_2}\delta_0 \ln\left|k_2(m^2 + n^2) - \frac{k_2}{k_1 - k_2}\delta_0\right|$$

$$= \frac{k_2}{k_2}(k_1 - k_2)m^2 + C_2, \tag{4.28}$$

when $k_1 \neq k_2$, where

$$k_1 = \frac{\partial V}{\partial X} \tag{4.29a}$$

and

$$k_2 = \frac{\partial U}{\partial Y}. \tag{4.29b}$$

The constants of integration C_1 and C_2 are determined by the initial condition of the wave packet.

4.4 Effect of the Topography

If we only consider the topography, the governing equations will become

$$\frac{D_g m}{DT} = \frac{1}{K^2}(mk_0 - n\mu), \tag{4.30}$$

$$\frac{D_g n}{DT} = -\frac{1}{K^2}(m\lambda + nk_0), \tag{4.31}$$

$$\frac{D_g}{DT}(m^2 + n^2) = \frac{2}{K^2}\{m^2 k_0 - n^2 k_0 + mn(\lambda + \mu)\}, \tag{4.32}$$

and

$$\frac{D_g}{DT}\left(-\frac{n}{m}\right) = \frac{1}{K^2}\left(-\lambda + \frac{2n}{m}k_0 - \frac{n^2}{m^2}\mu\right), \tag{4.33}$$

where

$$k_0 = \frac{\partial^2 \eta_B}{\partial y \partial X} = \frac{\partial^2 \eta_B}{\partial x \partial Y}, \tag{4.34a}$$

$$\mu = \frac{\partial^2 \eta_B}{\partial x \partial X}, \tag{4.34b}$$

and

$$\lambda = \delta_0 - \frac{\partial^2 \eta_B}{\partial y \partial Y}. \tag{4.34c}$$

From eqs. (4.30) to (4.34), it can be easily seen that the effect of a quadratic east–west-oriented topography upon the structural change of the Rossby wave packet only modifies the δ-effect. This is reminiscent of the β-effect, which is related to the propagation properties of the Rossby waves, where the linearly sloping topography only modifies the β-effect.

When the topography is in the case $k_0 = 0$ (e.g., symmetric topography), integrating along the characteristic line, that is, along the packet's group velocity, we obtain the following relation on the WKB phase plane of the packet:

$$\lambda m^2 - \mu n^2 = C, \tag{4.35}$$

where C is the integration constant to be determined by the initial condition.

Figure 4.10 illustrates the WKB trajectories of the evolution of a wave packet on such an axisymmetric topography. Figure 4.10a corresponds to the WKB trajectories on the β-plane, with $\mu = -0.5$, $\lambda = 0.5$, and $F = 1.0$. Here, if we consider the topography to be a elliptic surface, then the conditions that $\mu = -0.5$ and $\lambda = 0.5$ correspond to $\eta_B = H - 0.5xX - 0.5yY$. Figure 4.10b shows the results of the saddle topography on the β-plane. Figure 4.10a suggests that the packet in this case changes its whole spatial

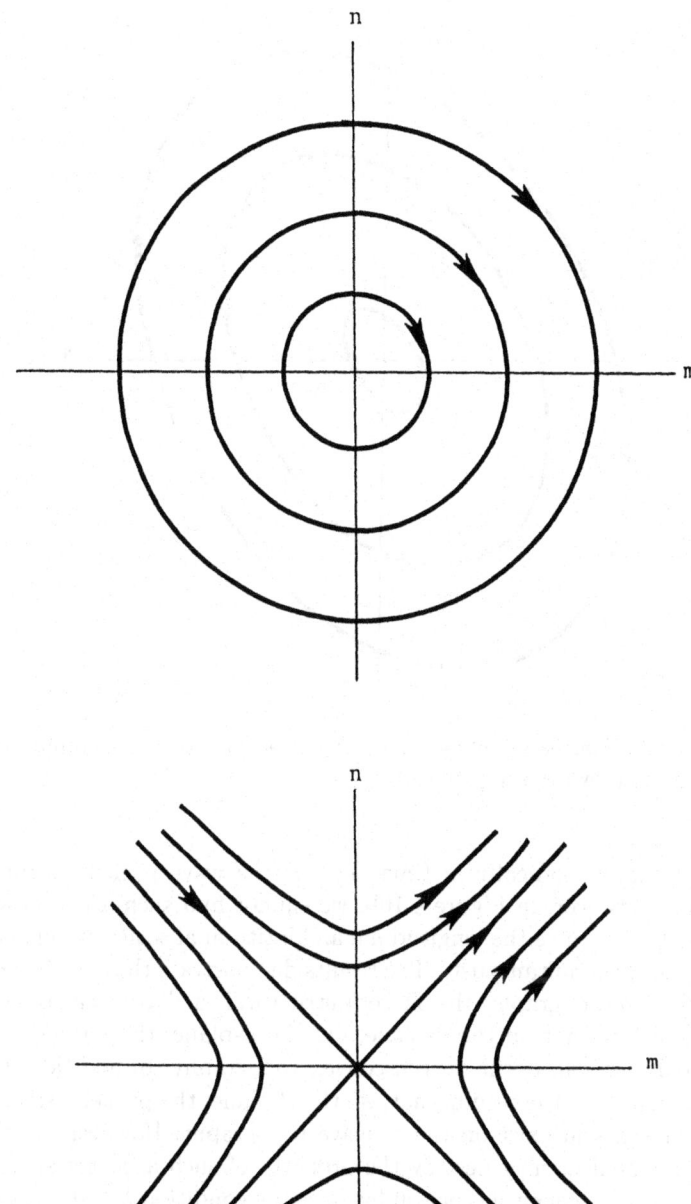

(a)

(b)

FIGURE 4.10. The WKB phase plane on the earth's β-plane in the presence of an axisymmetric topography. (a) On a convex topography and (b) on a saddle topography.

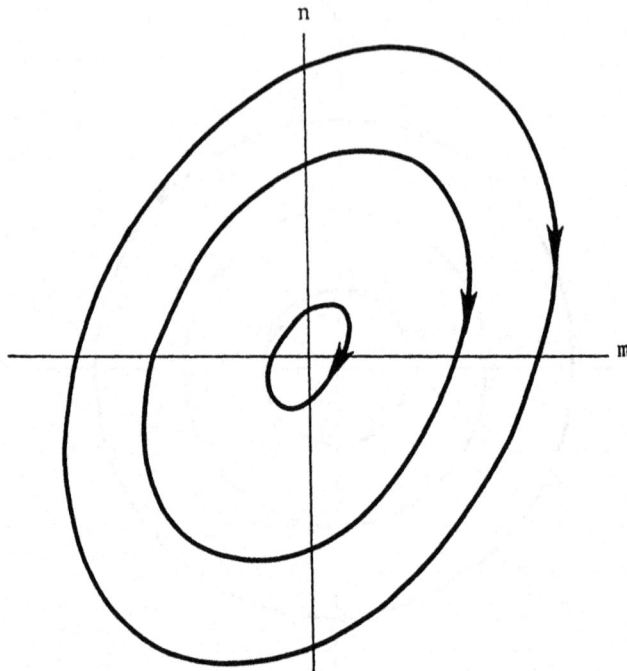

FIGURE 4.11. The WKP phase plane on an asymmetric topography on the earth's δ-surface. After Yang (1988a).

scale and tilts periodically in time. This is the wave packet's structural vacillation. However, in Figure 4.10b, we cannot find such a packet structural vacillation, since the longitudinal and latitudinal scales are increasing (or decreasing) simultaneously. The results demonstrate that on the north–south-oriented topography, the packet's structural vacillation can only exist on the δ-surface, not on the β-plane. On the β-plane, the packet's structural vacillation can occur on the convex topography. In addition, it can also be found from the results that, on the β-plane, the packet's structural vacillation can still exist on the concave topography. However, on the δ-surface, the δ-effect may destroy the property of such a packet structural vacillation on the concave topography, once the condition $\lambda > 0$ is satisfied. Moreover, in some cases on the saddle topography, the packet structural vacillation on the β-plane of the earth does not exist.

In the case $k_0 \neq 0$ (e.g., the asymmetric topography), we have to keep all the terms in (4.30) to (4.33). The asymmetry of the topography alters the evolution of the wave packet. The two governing systems, however, are dynamically similar.

Figure 4.11a shows the results of the WKB trajectories of the wave packet on the WKB phase plane in the case when $\mu = -0.5$, $\lambda = 0.5$, and $k_0 = -0.2$ on the β-plane. From the figure, one can readily find that the evolution on such an asymmetric topography differs from that on the axisymmetric topography. On the axisymmetric topography, the WKB trajectories of the wave packet on the WKB phase plane appear to be canonical ellipses. On the asymmetric topography, however, the trajectories are the rotated ellipses on the WKB phase plane, though the evolution of the wave packet still is time periodic. The difference means that on the symmetric topography, the wave packet's whole spatial scale will be largest or smallest when the packet is tilted parallel to the Y-axis, whereas on the asymmetric topography, the wave packet's whole scale will reach its maximum or minimum when the wave packet is tilted north–west to south–east or north–east to south–west, as, for example, in Figure 4.11. Figures 4.11a and b are WKB phase planes corresponding to the case on the β-plane and on the δ-surface, respectively. A comparison of Figure 4.11b with Figure 4.11a suggests that the δ-effect result in the trajectories rotating less.

In addition, by integrating the packet along the characteristic line, we obtain the complete integrals for the different cases, as shown in Appendix B.

4.5 Conclusions

From the above discussions, we can draw the following main conclusions:

1. The δ-surface approximation of the earth's surface can be used to describe structural change of a synoptic disturbance system. The presence of the δ-effect is of importance in the evolution of a Rossby wave packet and in the general dynamics of large-scale geophysical flows, especially near the earth's poles. The results show that the governing system on the earth's δ-surface might be dynamically different from that on the earth's β-plane. Therefore, on the δ-surface of the earth, the evolution of a Rossby wave packet will be different from that on the β-plane of the earth, either qualitatively or quantitatively. Our computations confirmed these arguments.

2. A wave packet's structural vacillation has been found on both the β-plane and the δ-surface in the presence of basic flow and/or topography. The wave packet's structural vacillation is characterized by the time periodic changes of the wave packet's structure. Both the tilt and the spatial scales (in the X-direction, in the Y-direction, and the whole) evolve toward periodic changes simultaneously. The wave packet's structural vacillation is characterized by the closed WKB trajectory on the WKB phase plane. The wave packet's vacillation suggests a possible mechanism of vacillations observed in the annulus and the atmosphere.

3. In the meridional basic flow or on the north–south-oriented topography, there is no such packet structural vacillation on the β-plane. On the δ-surface, however, there may be such packet structural vacillations.

4. The quadratic east–west-oriented topography only modifies the δ-effect in the structural change of the wave packet, which approximates the view that linearly sloping topography only modifies the β-effect on the propagation properties of Rossby waves.

5. The results show that the global behavior of the evolution of a Rossby wave packet may differ greatly. For example, in a southwesterly jet or in convex topographies, the WKB trajectories of the wave packet on the WKB phase plane appear to be closed lines, for example, ellipses, which suggests the existence of wave packet structural vacillations. In a southeasterly jet or on saddle topographies, for example, the WKB trajectories on the WKB phase plane appear to be open lines, for example, hyperbola or straight lines, which suggests that both the spatial scales of the wave packet in the X-direction and in the Y-direction are increasing (or decreasing) simultaneously.

The wave packet's structural vacillation differs somewhat from wave vacillations observed in the annulus and the atmosphere as, for example, tilt troughline vacillation (e.g., Hide, 1958), amplitude vacillation (e.g., Pfeffer and Chiang, 1967), and structural vacillation (e.g., Pfeffer et al., 1980a,b; Buzyna et al., 1984), since the wave packet's structural vacillation evolves a time periodic change in both the tilt and the spatial scales (in the X-direction, in the Y-direction, and the whole) simultaneously. Therefore, the process of vacillations in the annulus and in real geophysical flows (e.g., Gruber, 1975; Webster and Keller, 1975; McGuirk and Reiter, 1976), will be much more complicated than the packet structural vacillation found in this study. Yet, if one carefully reexamines the time series of the wave vacillations in the experiment, for example, as shown in Figure 4.5, the spatial scales of the wave are actually changing as the tilt changes with time. Understanding the mechanism for such wave vacillations in the annulus and real geophysical flows is still challenging investigators. It is surprising to notice that, though only a simple barotropic model was employed, such wave packet structural vacillations could, very neatly, still be found, since generally, the wave structural vacillations have been mainly considered baroclinic processes. The role of topography in the atmosphere has attracted increasing attention recently (e.g., Hart, 1979; Charney and DeVore, 1979; Hendon, 1986). The present results suggest a new role of the topography in geophysical flows. Furthermore, the results strongly suggest that the topography and the basic flow could produce wave packet structural vacillation in the β-plane and in the δ-surface. In spite of this, we still should be careful when we to apply the present theory directly to real geophysical flows and the annulus, since there the baroclinic process may dominate in some cases and the flows will be more complicated than in the present model. Obviously, further work is needed to test the prediction by using real data and making observations.

Appendix A. The Wave Packet Solutions for the Asymmetric Basic Flow on the Earth's β-Plane

Since there is a relationship between m and n, that is, (3.9), we only need give the expressions for m. For example, we only consider the case that $(\partial U/\partial Y)(\partial V/\partial X) < 0$ and $(\partial V/\partial X) > 0$.

Define

$$k^2 = -\frac{\partial U}{\partial Y}\frac{\partial V}{\partial X} > 0. \tag{4.36}$$

Then, when $1 - k^2 = 0$,

$$\left(1 + \frac{F}{C_1}\sin\frac{mk}{\sqrt{C}}\right) = -\frac{\partial V}{\partial X}T + C_2; \tag{4.37}$$

otherwise,

$$\arcsin\frac{mk}{\sqrt{C_3}} + \frac{F}{(1-k^2)\left(\frac{C_3}{1-k^2}\right)^{\frac{1}{2}}\left(\frac{C_3}{1-k^2} + \frac{C_3}{k^2}\right)^{\frac{1}{2}}}$$

$$\times \arctan\frac{m\left(\frac{C_3}{1-k^2} + \frac{C_3}{k^2}\right)^{\frac{1}{2}}}{\left(\frac{C_3}{1-k^2}\right)^{\frac{1}{2}}\left(\frac{C_3}{k^2} - m^2\right)^{\frac{1}{2}}} = -\frac{\partial V}{\partial X}kT + C_4. \tag{4.38}$$

Here, C_1, C_2, C_3, and C_4 are constants to be determined by the initial condition and $C = C_1 k^2$.

Appendix B. The Expressions in the WKB Phase Space for the Asymmetric Topography on the Earth's δ-Surface

When $\lambda = 0$,

$$m = -\frac{\mu}{k_0}n\ln n + C_1 n \tag{4.39a}$$

or

$$n = C_1' e^{-(k_0/\mu)(m/n)}. \tag{4.39b}$$

When $\mu\lambda > 0$,

$$\frac{1}{2}\ln\left(\frac{m^2}{n^2} + \frac{\mu}{\lambda}\right) + \frac{k_0}{\lambda}\arctan\left(\left(\frac{\lambda}{\mu}\right)^{\frac{1}{2}}\frac{m}{n}\right) = \ln n + C_2 \quad (4.40a)$$

or

$$\left(m^2 + \frac{\mu}{\lambda}n^2\right)^{\frac{1}{2}} = C_2' n^2 e^{-[k_0/\sqrt{\mu\lambda}]\arctan\{[\sqrt{\lambda/\mu}](m/n)\}}. \quad (4.40b)$$

When $\mu\lambda < 0$,

$$\ln\left(m^2 + \frac{k_2}{\lambda}n^2\right)^{\frac{1}{2}} + \frac{k_0}{2\sqrt{|\mu\lambda|}}\times\ln\frac{m - |\lambda/\mu|^{\frac{1}{2}}}{m + |\lambda/\mu|^{\frac{1}{2}}} = \ln n^2 + C_3, \quad (4.41a)$$

or

$$\left(m^2 + \frac{\mu}{\lambda}\right)^{\frac{1}{2}}\left(\frac{m - |\lambda/\mu|^{\frac{1}{2}}}{m + |\lambda/\mu|^{\frac{1}{2}}}\right)^{k_0/[2\sqrt{|\mu\lambda|}]} = C_3' n^2. \quad (4.41b)$$

Here, C_1, C_2, C_3, C_1', C_2', and C_3' are integration constants to be determined by the initial condition.

REFERENCES

Buzyna, G., Pfeffer, R.L., and Kung. R. (1984). Transition to geostrophic turbulence in a rotating differentially heated annulus of fluid. *J. Fluid Mech.* **145**, 377–403.

Charney, J.G., and DeVore, J.G. (1979). Multiple flow equilibria in the atmosphere and blocking. *J. Atmos. Sci.* **36**, 1205–1216.

Elberry, R.L. (1968). A high-rotating general circulation model experiment with cyclic time changes. Atmospheric Science Paper No. 134, Colorado State University, Fort Collins.

Fultz, D., Long, R.R., Owens, G.V., Bohan, W., Kaylor, R., and Weil, J. (1959). Studies of thermal convection in a rotating cylinder with some implications for large-scale atmospheric motions. *Meteorol. Monogr.* **21**, American Meteorol. Soc., Boston.

Gruber, A. (1975). The wave-number-frequency spectra of the 200 mb wind field in the tropics. *J. Atmos. Sci.* **32**, 1283–1300.

Hart, J.E. (1979). Barotropic quasi-geostrophic flow over anisotropic mountains. *J. Atmos. Sci.* **36**, 1736–1746.

Hendon, H.H. (1986). Time-mean flow and variability in a nonlinear model of the atmosphere with orographic forcing. *J. Atmos. Sci.* **43**, 433–448.

Hide, R. (1958). An experimental study of thermal convection in a rotating liquid. *Phil. Trans. Roy. Soc. London* **A250**, 441–478.

Karoly, D.J. (1983). Rossby wave propagation in a barotropic atmosphere. *Dynamics Atmos. Oceans* **7**, 111–125.

McGuirk, J.P., and Reiter, E.R. (1976). A vacillation in atmospheric energy parameters. *J. Atmos. Sci.* **33**, 2079–2093.

Pfeffer, R.L., and Chiang, Y. (1967). Two kinds of vacillation in rotating laboratory experiments. *Month. Weath. Rev.* **95**, 75–82.

Pfeffer, R.L., Buzyna, G., and Fowlis, W.W. (1974). Synoptic features and energetics of wave-amplitude vacillation in a rotating, differentially heated fluid. *J. Atmos. Sci.* **31**, 622–645.

Pfeffer, R.L., Buzyna, G., and Kung, R. (1980a). Time-dependent modes of behavior of thermally driven rotating fluids. *J. Atmos. Sci.* **37**, 2129–2149.

Pfeffer, R.L., Buzyna, G., and Kung, R. (1980b). Relationships among eddy fluxes of heat, eddy temperature variances and basic-state temperature parameters in thermal driven rotating fluids. *J. Atmos. Sci.* **37**, 2577–2599.

Webster, P.J., and Keller, J.L. (1975). Atmospheric variations: Vacillations and index cycles. *J. Atmos. Sci.* **32**, 1283–1300.

Yang, H. (1987). Evolution of a Rossby wave packet in barotropic flows with asymmetric basic flow, topography and δ-effect. *J. Atmos. Sci.* **44**, 2267–2276.

Yang, H. (1988a). Global behavior of the evolution of a Rossby wave packet in barotropic flows on the earth's δ-surface. *J. Atmos. Sci.* **45**, 134–146.

Yang, H. (1988b). Bifurcation properties of the evolution of a Rossby wave packet in barotropic basic flows on the earth's δ-surface. *J. Atmos. Sci.* **45**, 3667–3683.

Yang, H. (1988c). Secondary bifurcation of the evolution of a Rossby wave packet in barotropic flows on the earth's δ-surface. *J. Atmos. Sci.* **45**, 3684–3699.

Zeng, Q.-C. (1982). On the evolution and interaction of disturbances and zonal flow in rotating barotropic atmosphere. *J. Meteorol. Soc. Japan* **60**, 24–31.

5

Change in Global Behavior: Bifurcation

5.1 Introduction

In earlier chapters, we mainly discussed the effects of basic flow and topography separately. In real geophysical flows, however, both factors are often of importance, especially for low level flows. Therefore, it is worthwhile to investigate the problem from the point of view further, in order to better understand the roles of topography and basic flow. In what follows, we combine these two factors to investigate the bifurcation properties of the evolution of a wave packet (Yang, 1988b).

Charney (Charney and DeVore, 1979) initiated the multiple equilibria, Vickroy and Dutton (1979) investigated bifurcation and catastrophe, and Winn-Nielsen (1979) studied steady states and their stabilities in a sphere. Since then, the bifurcation properties in the geophysical fluids have attracted increasing attention. In fact, the study of bifurcation properties in geophysical flows goes back to Lorenz (1963), whose work in dynamics has been extensively studied (e.g., Curry, 1978; Sparrow, 1982). It is believed that in geophysical fluids, the topography plays a very important role in bifurcation properties. Källen (1982) studied a low-order barotropic system with topographic and momentum forcing on a sphere. He showed that low-order model, multiple equilibria develop as a result of a sufficiently strong, long-wave, orographic forcing and that a suitable positioned wave vorticity forcing can enhance the bifurcation. Mitchell and Dutton (1981), however, examined a forced, dissipative barotropic model on a cyclic β-plane and found a Hopf bifurcation at a certain critical value of the forcing. They also found that the degree of nonlinearity, scale of the forcing, the spatial dependence of the disturbance, and the forcing all crucially influence both the multiplicity and the temporal nature of the stable limit solutions in their model. Using a numerical model, Yoden (1985) found several new bifurcation properties. Moroz and Holmes (1984), from a different point of view, analytically and numerically investigated double Hopf bifurcation and quasi-periodic properties near the flow's baroclinic instability. The work done by Tung and Rosenthal (1985) is also noteworthy.

In this chapter, we analytically consider the bifurcation properties of a barotropic model from a different point of view, namely, the structural transitions of a wave packet. It should be noted, however, that the theory of bifurcations generally is concerned with the changes that occur in the topological structure of a dynamic system in a particular region, when the system itself is altered, and the term *bifurcation* generally refers to these changes in topological structure. In the following, it is shown that the topological structure of the evolution of a wave packet in various basic flows changes with the topography parameter accordingly, in the present model. The *homoclinic orbits* and *heteroclinic orbits* are also discussed; these orbits are separatrices of the different kinds of the wave packet's structural vacillations (oscillations) found in earlier chapters.

The material in this chapter is arranged as follows: The basic theory of the dynamic system is addressed in the following section; it includes stability theory and basic bifurcation theory. The main analysis of the bifurcation properties on symmetric and asymmetric topography with a symmetric basic flow is presented in Section 5.3 and Section 5.4, respectively. Implications of the different kinds of bifurcations in the evolution of a wave packet are discussed in Section 5.5, using the WKB phase space. The chapter ends with a summary.

5.2   Basic Theory of the Dynamic System

In this section, we briefly review some results in the theory of the dynamic system, especially the stability and the bifurcation theory. The bifurcation theory is always associated with equilibrium solutions and the stabilities of their dynamic system; thus, the equilibrium solution of the dynamic system will be discussed first, and the basic stability theory of the dynamic system second. Many authors have discussed the stability theory of the dynamic system. However, this author found that the book by Hirsh and Smale (1974) is an excellent introductory textbook, and highly recommends it to those who are not very familiar with the theory, since it provides not only basic analytic theorems but also geometric illustrations, which are easily understood. After a discussion of stability theory and topological structures near equilibria, the bifurcation theory of the dynamic system is discussed. Iooss and Joseph (1980) discuss the bifurcation theory from an analytic point of view. Moreover, the books by Golubitsky and Schaeffer (1985) and by Golubitsky, Stewart, and Schaeffer (1988) approach the bifurcation theory from a singularities and groups point of view, in the belief that singularity theory offers an extremely useful approach to bifurcation problems. *The Theory of Bifurcations of Dynamical Systems on a Plane* presented the early work of bifurcation theory by the Russian (Andronov et al. 1971; Russian original in 1967), which provides systematic theorems and some detailed examples of bifurcation on a two-dimensional dynamic system. However, the book by Guckenheimer and Holmes (1983) is a more advanced consideration of the theory, both analytically and geometrically.

5.2.1 EQUILIBRIA AND BASIC STABILITY THEORY

The dynamic system, as in (4.1) and (4.2), can be written

$$\dot{\mathbf{X}} = \mathbf{F}(\mathbf{X}, \lambda, t), \tag{5.1}$$

where $\mathbf{X}$ is a vector of variables, such as m and n in (4.1) and (4.2), and the dot denotes the material derivative of the vector with respect to time t. Also $\mathbf{F}$ is the field vector of the dynamic system and λ is a parameter, or in general, a parameter set, of the dynamic system. If the field vector $\mathbf{F}$ does not *explicitly* depend upon time, then the dynamic system (5.1) can be written as

$$\dot{\mathbf{X}} = \mathbf{F}(\mathbf{X}, \lambda) \tag{5.2}$$

and is said to be *autonomous*. The dynamic system (5.1) in which the field vector $\mathbf{F}$ explicitly depends on time is said to be *nonautonomous*. A nonautonomous dynamic system can always be transformed into an autonomous dynamic system by increasing it by one degree of freedom. Therefore, in what follows, we only consider the autonomous dynamic system.

The equilibrium states of the dynamic system (5.2) are defined as the solution satisfying

$$\mathbf{F}(\mathbf{X_e}, \lambda) = 0. \tag{5.3}$$

Various names can be found in literature, such as *steady states, fixed points, stationary states, stationary solutions, steady points,* or *nonwandering points.*

The equilibrium state can be either stable or unstable depending on one's intuitive feeling. Every one has an intuitive feeling about what stability or instability means. The mathematics concepts, however, is very subtle, and rigorous definitions are necessary, though a theorem stated by Lagrange and proved by Dirichlet assures that an equilbrium state or position of the (conservative) system (viz, a time-independent solution of Lagrange differential equations) is stable if the potential energy has a minimum there. The precise mathematical definitions for stability theory had not been established until Lyapunov's 1893 work. Lyapunov also proved under restrictions that the same position is unstable if the potential energy has no minimum here, or if it is a maximum. This theory is called the *Lyapunov stability theory.*

In the following, we define stability in the sense of Lyapunov's stability and restrict our consideration to the null solution of the system (5.2) as the equilibrium state without loss of generality. If the equilibrium state is not located at the origin, a displacement transformation suffices to relocate it there.

(1) Lyapunov First Method

Stability. Let $\mathbf{X}(t)$ be a solution of (5.2). We say that $\mathbf{X}(t)$ is stable if,

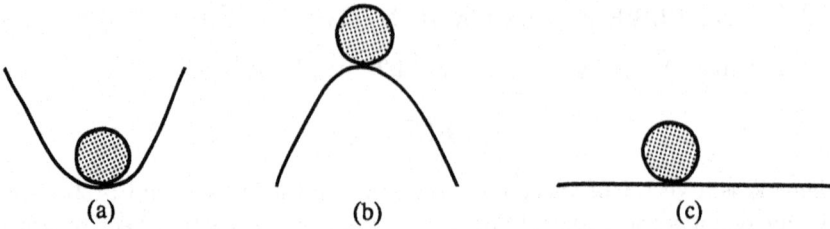

FIGURE 5.1. (a) Stable, (b) unstable, and (c) asymptotically stable mechanical equilibrium states of a billiard ball on a perfectly smooth surface.

given $\varepsilon > 0$ and t_0,

$$\eta = \eta(\varepsilon, t_0),$$

such that any solution $\mathbf{Y}(t)$ for which

$$\|\mathbf{X}(t_0) - \mathbf{Y}(t_0)\| < \eta \tag{5.4}$$

satisfies

$$\|\mathbf{X}(t) - \mathbf{Y}(t)\| < \varepsilon, \tag{5.5}$$

for $t > t_0$. If η may be chosen independently of t_0, $\mathbf{X}(t)$ is said to be *uniformly stable*. If no such η exists, $\mathbf{X}(t)$ is *unstable*.

Asymptotic stability. If $\mathbf{X}(t)$ is stable, and in addition,

$$\|\mathbf{X}(t) - \mathbf{Y}(t)\| \to 0, \quad \text{as } t \to +\infty, \tag{5.6}$$

we say that it is *asymptotically stable*. An asymptotically stable, equilibrium state or trajectory is usually called *an attractor of the motion*.

It should be pointed out that both of these definitions concern solutions (motions). Intuitively they can be taken to mean that a solution (or motion) is stable if all solutions coming near it remain in its neighborhood; whereas it is asymptotically stable if the solution approaches it as time increases (and eventually goes to infinity). Figure 5.1 illustrates these concepts by means of the possible motions of, say, a billiard ball on a perfectly smooth surface under the influence of the gravitational field. For Figures 5.1a and 5.1c, the equilibrium position is stable; whereas for Figure 5.1b, it is unstable. Moreover, in Figure 5.1a the equilibrium position is not only stable but also asymptotically stable; whereas in Figure 5.1c, it is stable but not asymptotically stable.

According to these definitions, the periodic motions are stable in the sense of Lyapunov only when there is *isochronism*. However, even in this case, the motions are not asymptotically stable. Therefore, we need to introduce the following definition of stability for periodic motion.

Orbital stability. In general, a sufficient close path is said to be orbitally stable if it will always lie in its entirety in the immediate vicinity of the chosen one; otherwise, it is said to be orbitally unstable.

(2) Lyapunov's Second Method or Lyapunov's Direct Method

However, it is not always possible to have explicit knowledge of the solution of the dynamic system. Lyapunov's direct method provides a way to determine stability directly from the dynamic system without any knowledge of its solution. Here we summarize a few theorems of the stability theory by use of this approach.

Let $V(\mathbf{X})$ be a scalar function of $\mathbf{X}$, and W be an open domain about the origin $(X = 0)$.

Positive definition of functions. The function $V(\mathbf{X})$ is positive definite in W, if for all $\mathbf{X}$ and W

   (i) it is continuous together with its first partial derivatives;

   (ii) it vanishes at the origin; and

   (iii) it is positive outside the origin (and always in W). The origin is an isolated minimum.

Note: If in the condition (iii) we replace the term positive by negative, we have a *negative definite function*. If, however, we relax condition (iii) to allow the function to be positive or to vanish, this function is said to be *positive semidefinite*. Similarly, we might have a *negative semidefinite function*.

(3) Lyapunov Theorem I: Stability

The null solution, or the equilibrium state at the origin of (5.2), is stable if there is some neighborhood of the origin where a positive definite function $V(X)$ exists such that its gradient $\dot{V}(X)$ with respect to the solutions of (5.2) is negative semidefinite, that is,

$$\dot{V}(\mathbf{X}) \leq 0. \tag{5.7}$$

(4) Lyapunov Theorem II: Asymptotic Stability

The null solution of system (5.2) is asymptotically stable if, in some neighborhood of the origin, there is a positive definite function $V(X)$ such that its gradient $\dot{V}(X)$ with respect to the solutions of (5.2) is negative, that is,

$$\dot{V}(\mathbf{X}) < 0. \tag{5.8}$$

A function $V(X)$ satisfying either theorem is called a *Lyapunov function*. Clearly, for a given system there are as many Lyapunov functions as one may be able to find, in principle, an infinite number of them. It should be noted that two different Lyapunov functions may yield two different estimates for the region (extent) of the asymptotic stability. There are, however, no general methods for finding suitable Lyapunov functions, although in mechanical problems, energy is often a good candidate.

When the whole space is the region of asymptotic stability, we have global complete asymptotically stability. The book by Rouche, Habets, and Laloy (1977) gives a complete account of the development of the theory from a mathematics point of view. An extensive application of Lyapunov theory in hydrodynamics can be found in the book by Joseph (1976).

5.2.2   TOPOLOGICAL STRUCTURE NEAR EQUILIBRIA

We are now going to look at the topological structure near an equilibrium state. The way to test stability is to introduce small variables at each equilbrium state, that is, to perturb the equilibrium state, linearize the equations, and compute the stability factor (eigenvalues). If the real parts of the eigenvalues are all negative, the equilibrium state is stable. If any one of the real parts of the eigenvalues is positive, the equilibrium state is unstable. We illustrate the topological structure near equilibrium on a two-dimensional space. The set of the two-dimensional dynamic system is

$$\frac{dx}{dt} = f(x, y) \tag{5.9a}$$

and

$$\frac{dy}{dt} = g(x, y). \tag{5.9b}$$

Introducing small variables about equilibrium state (x_e, y_e), that is,

$$x = x_e + \eta \quad \text{and} \quad y = y_e + \xi, \tag{5.10}$$

we can write the linearized system in the form

$$\frac{d}{dt}\begin{pmatrix} \eta \\ \xi \end{pmatrix} = \begin{pmatrix} \frac{\partial f}{\partial x} & \frac{\partial f}{\partial y} \\ \frac{\partial g}{\partial x} & \frac{\partial g}{\partial y} \end{pmatrix} \begin{pmatrix} \eta \\ \xi \end{pmatrix}, \tag{5.11}$$

where the derivatives are evaluated at the point (x_e, y_e). The nature of the motion about each equilibrium state is determined by looking for eigensolutions

$$\begin{pmatrix} \eta \\ \xi \end{pmatrix} = \begin{pmatrix} \alpha \\ \beta \end{pmatrix} e^{pt}, \tag{5.12}$$

where α and β are constants and p is the stability factor. The motion is classified according to the nature of the stability factors, that is, eigenvalues of $\nabla \mathbf{F}$ where $\mathbf{F} = (f, g)$ and represents the matrix of the partial derivatives.

The stability of the linearized system (5.11) depends on the sign of Real (p). When one of the real parts of p_1 and p_2 is positive, the motion about the equilibrium state is unstable. If all the real parts are negative, then the

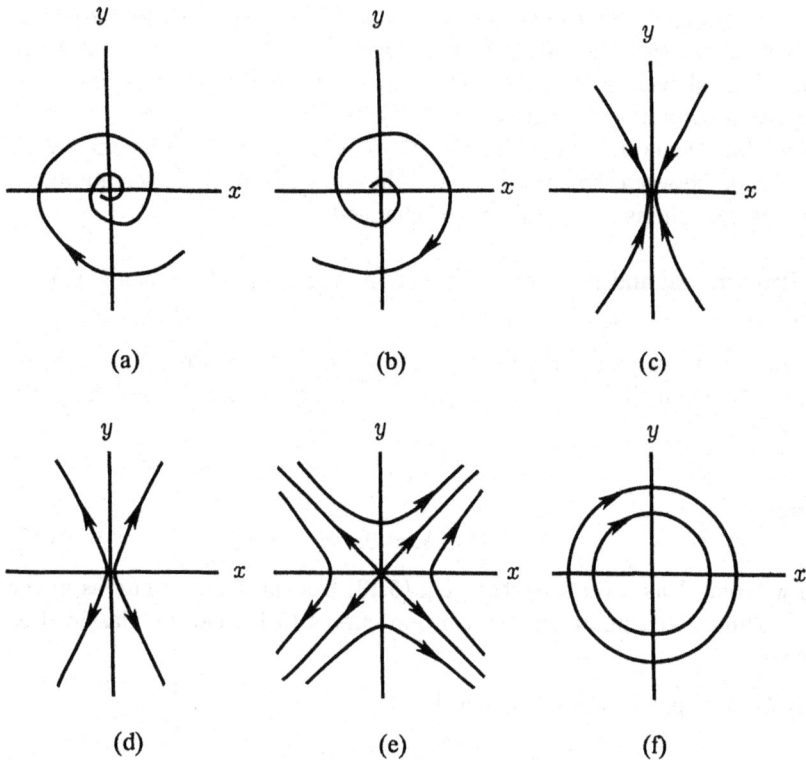

FIGURE 5.2. Classical phase plane portraits near different types of equilibrium states for an autonomous two-dimensional dynamic system. (a) Stable foci, (b) unstable foci, (c) stable node, (d) unstable node, (e) saddle, and (f) center.

motion is asymptotically stable. Sketches of trajectories in the phase plane for different eigenvalues are shown in Figure 5.2. For example, the saddle point is obtained when both eigenvalues p_1 and p_2 are real, but $p_1 p_2 < 0$. A spiral appears when p_1 and p_2 are complex conjugates. A node appears when both eigenvalues p_1 and p_2 are real and $p_1 p_2 > 0$. A center exists when the eigenvalues are purely imaginary.

5.2.3 BIFURCATION THEORY

The bifurcation theory is concerned with the change that occurs in the topological structure of a dynamic system in a particular region when the system itself is altered by means of a change in the system parameter, and the term *bifurcation* originally refers to these changes in topological structure. Generally speaking, the stability of equilbrium states in a dynamic system can change as well as the number of equilibrium states, as the parameter of the dynamic system is changed. The study of these changes in nonlinear problems as the system parameter is changed is the subject of bifurcation theory. The values of the parameters at which the qualitative or topological nature of motion changes are known as *critical values* or *bifurcation values*. The term *bifurcation point* is also often used.

Therefore generally speaking, the bifurcation theory is concerned with the relation between the parameter and the equilibrium states of a nonlinear dynamic system, as the parameter is changed.

(1) Transcritical and Pitchfork Bifurcation, and Bifurcation at a Cusp Point

For simplicity, we consider a one-dimensional dynamic system, in which the relation between the parameter and the equilibrium state (5.3) is reduced to the form

$$x_e = \varepsilon,$$

where

$$F(\varepsilon, \lambda) = 0. \tag{5.13}$$

If we treat λ as a variable, then eq. (5.13) has the solution curves in the (ε, λ) plane. Each point on the solution curves (5.13) can be classified as follows:

(i) *Regular point:* a point at which

$$F_\varepsilon \neq 0 \quad \text{or} \quad F_\lambda \neq 0, \tag{5.14}$$

where

$$\lambda = \lambda(\varepsilon) \quad \text{or} \quad \varepsilon = \varepsilon(\lambda). \tag{5.15}$$

In this case, we can find a unique curve.

(ii) *Regular turning point:* a point at which λ_ε changes sign and

$$F_\lambda(\varepsilon, \lambda) \neq 0. \tag{5.16}$$

(iii) *Singular point:* a point at which

$$F_\lambda = F_\varepsilon = 0. \tag{5.17}$$

(iv) *Double point:* the point through which pass two and only two branches of $F(\varepsilon, \lambda) = 0$ possessing distinct tangents, when we assume that all second derivatives of F do not simultaneously vanish at that point.

(v) *Singular turning (double) point:* a double point at which λ_ε changes sign on one branch.

(vi) *Cusp point:* the point of second order contact between two branches of the curves. The two branches of the curve have the same tangent at a cusp point.

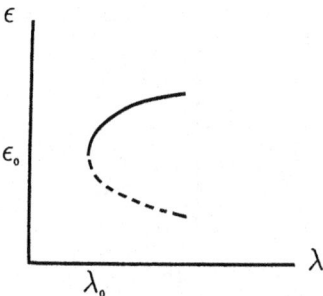

FIGURE 5.3. The saddle-node bifurcation near a turning point. Key: - - -, unstable equilibrium state; —, stable equilibrium state.

(vii) *Conjugate point:* an isolated singular point solution of $F(\varepsilon, \lambda) = 0$.

(viii) *Higher-order singular point:* a singular point at which all three second derivatives of $F(\varepsilon, \lambda)$ vanish.

The factorization theorem in one-dimensional states is as follows: For every equilibrium solution $F(\lambda, \varepsilon) = 0$ for which $\lambda = \lambda(\varepsilon)$, we have

$$p(\varepsilon) \equiv F_\varepsilon(\lambda(\varepsilon), \varepsilon) = -\lambda_\varepsilon(\varepsilon) F_\lambda(\lambda(\varepsilon)) \stackrel{\text{def}}{=} -\lambda_\varepsilon \hat{p}(\varepsilon). \qquad (5.18)$$

This theorem is easy to prove. Since $\lambda = \lambda(\varepsilon)$,

$$F(\lambda, \varepsilon) = F(\lambda(\varepsilon), \varepsilon) = 0$$

and

$$\frac{dF}{d\varepsilon} = F_\varepsilon(\lambda(\varepsilon), \varepsilon) + \lambda_\varepsilon F_\lambda(\lambda(\varepsilon), \varepsilon) = 0.$$

Therefore,

$$F_\varepsilon(\lambda(\varepsilon), \varepsilon) = -\lambda_\varepsilon(\varepsilon) F_\lambda(\lambda(\varepsilon), \varepsilon). \qquad (5.19)$$

Here $p(\varepsilon)$ must change sign as ε is varied across a regular turning point. This implies that the solution $x_e = \varepsilon$, $\lambda = \lambda(\varepsilon)$ is stable on one side of a regular turning point and is unstable on the other side. Figure 5.3 illustrates the stability of an equilibrium solution near a turning point. This is called a *saddle-node bifurcation.*

If all singular points of solution of $F(\lambda, \varepsilon) = 0$ are double points, the stability of such solutions must change at each regular turning point and at each singular point (which is not a turning point), and only at such points. At a double point, it can be shown that there are three possible bifurcations: *transcritical bifurcation, supercritical bifurcation,* and *subcritical bifurcation.* The transcritical bifurcation is also called a *two-sided bifurcation.* Bifurcation diagrams near a double point are given in Figure 5.4. These supercritical and subcritical bifurcations in Figure 5.4b and Figure 5.4c, are also called *pitchfork bifurcations.*

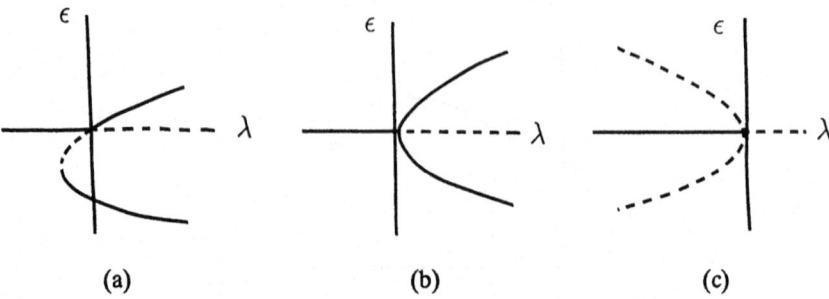

FIGURE 5.4. Bifurcation diagrams near a double point. The bifurcation value is zero. (a) Transcritical bifurcation, (b) supercritical bifurcation, and (c) subcritical bifurcation. Key: - - -, unstable equilibrium state; —, stable equilibrium state.

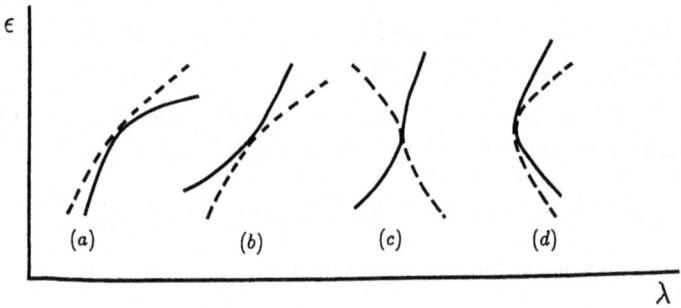

FIGURE 5.5. Bifurcation at a cusp point.

The bifurcation at a cusp point is given in Figure 5.5. It can be shown that the stability of any branch passing through a cusp point of second order changes sign if and only if $\lambda_\varepsilon(\varepsilon)$ does.

The *Duffing oscillator* is a good example of a bifurcation in a two-dimensional dynamic system, which is described as follows:

$$\frac{dx}{dy} = y \tag{5.20}$$

and

$$\frac{dy}{dt} = -(\lambda x + \mu x^3). \tag{5.21}$$

Physically, the force $-(\lambda x + \mu x^3)$ can be derived from a potential energy function. The positive μ and the negative μ correspond to, respectively, the hard spring problem and the soft spring problem. The bifurcation diagram for the hard spring problem can be obtained as shown in Figure 5.6, which is the reverse of that in Figure 5.4b when λ is taken as the bifurcation

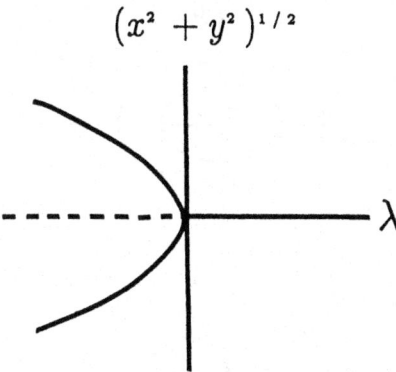

FIGURE 5.6. Bifurcation diagram of a hard spring problem, a reverse supercritical bifurcation.

parameter. Therefore, this is the *reverse supercritical bifurcation*. As the bifurcation parameter changes from positive to negative, one equilibrium point splits into three points. When the bifurcation parameter becomes negative; a one-well potential changes into a double-well potential problem. Dynamically, one center is transformed into a saddle point at the origin and two centers. This represents a qualitative change in the dynamics, and thus $\lambda = 0$ is a critical bifurcation value. Figure 5.7 shows the phase space for the hard spring problem in the Duffing oscillator when the bifurcation parameter is positive in Figure 5.7a and negative in Figure 5.7b, and for the soft spring problem in the Duffing oscillator in Figure 5.7c.

(2) The Hopf Bifurcation

The *Hopf bifurcation* is the bifurcation that occurs when the periodic solution is being bifurcated from a steady solution as the bifurcation parameter reaches a critical value. The theory was originated by Poincaré (1892) and was extensively developed by Andronov and Witt (1930) and their colleagues. Hopf (1942) extended the theory from two dimensions to higher dimensions. Readers who are interested in this subject are encouraged to consult the book by Marsden and McCracken (1976), in which the English translation of Hopf's original paper and excellent comments on the paper by L.N. Howard are available.

Suppose that $\mathbf{u} = \mathbf{0}$ is the steady state for all range of bifurcation parameter λ. The eigenvalues of a linearized system about the steady state are

$$p = p_r(\lambda) + ip_i(\lambda). \tag{5.22}$$

If

(i) $p_r(\lambda_c) = 0$, at $\lambda = \lambda_c$;

(ii) $\omega_0 = p_i(\lambda_c) \neq 0$; and

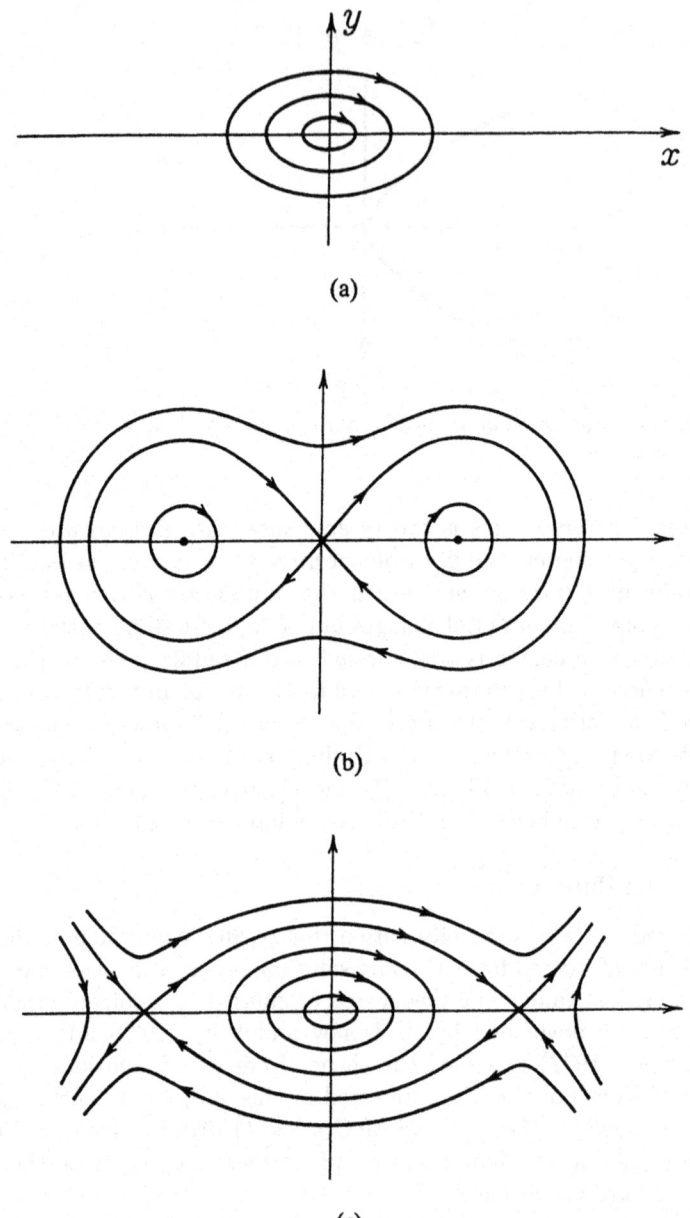

(a)

(b)

(c)

FIGURE 5.7. Phase plane diagrams for the Duffing oscillator. (a) Hard spring problem, $\lambda > 0$, $\mu > 0$; (b) two-well potential, $\lambda < 0$, $\mu > 0$; and (c) soft spring problem, $\lambda > 0$, $\mu < 0$.

(iii) $\frac{dp_r}{d\lambda}\big|_{\lambda=\lambda_c} \neq 0$; then, the steady solution will be bifurcated into peri-
odic solutions at $\lambda = \lambda_c$. The λ_c is the bifurcation critical value of
the Hopf bifurcation.

The following example shows this kind of bifurcation. Consider a two-
dimensional dynamic system such as

$$\frac{dx}{dt} = \lambda x - y - x^3 \tag{5.23}$$

and

$$\frac{dy}{dt} = x + \lambda y - y^3, \tag{5.24}$$

which is similar to the one in the previous discussion. The system has a
equilibrium state $x = y = 0$ for all values of the bifurcation parameter λ. It
can be shown analytically that this solution is stable for $\lambda < 0$ and unstable
for $\lambda > 0$. Moreover, a periodic solution is bifurcated from the steady state
at $\lambda = 0$.

Linearizing the dynamic system (5.23) and (5.24) about the equilibrium
state, $x = y = 0$, yields the stability factor as

$$p = \lambda \pm i, \tag{5.25}$$

so that the equilibrium state, $x = y = 0$, is stable for $\lambda < 0$ and unstable
for $\lambda > 0$. Furthermore, the point $x = y = 0$ in the phase plane is a stable
focus for $\lambda < 0$ and an unstable focus for $\lambda > 0$. By the use of the regular
perturbation method to the third order, it can easily be proven that there
appears a periodic solution near the origin when the parameter λ is greater
than zero. Moreover, it can be shown that the periodic solution only exists
and is stable when the parameter λ is greater than zero. The period of
the solution depends on the initial amplitude of the first order solution.
Therefore, we conclude that there is a Hopf bifurcation at $\lambda = 0$, as shown
in Figure 5.8. The Hopf bifurcation will not occur in a one-dimensional
autonomous dynamic system.

5.2.4 STRUCTURAL STABILITY

A dynamic system (A) is said to be *structurally stable* if there is an ε such
that all C^1, ε perturbations of the system are topologically equivalent to
the system (A). Therefore, if the topological structure of the perturbation
system is the same as that of the original system, the system is structurally
stable; otherwise, it is *structurally unstable*.

If system (A) is structurally stable in W, the topological structure of the
partition of some neighborhood H of W by the paths of system (A) does
not change in a certain sense on passing to a sufficiently close system $(\tilde{A})$,
or precisely, an infinitesimal translation will transform H into $\tilde{H}$ so that the

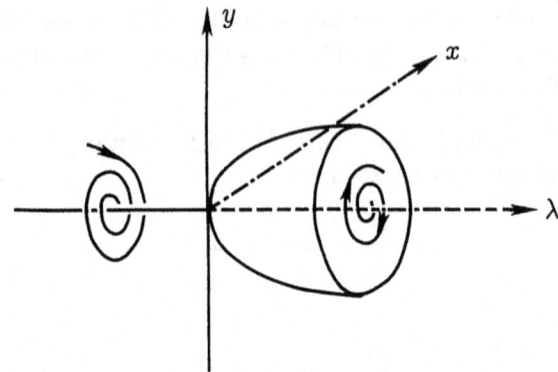

FIGURE 5.8. The Hopf bifurcation.

paths of (A) coincide with the paths of ($\tilde{A}$). This property explains the term *structural stability*. An alternative term used in the Russian literature is *coarse system*, which implies that the topological structure of the partition of a given region by paths is not affected by small changes in the system (A), or in other words, the structure can resist small disturbances in the system (A).

Structurally stable systems were first considered in 1937 by Andronov and Pontryagin, who originally called their systems *grossier,* or *coarse systems.*

In particular, for a two-dimensional system, we have the following theorem about the structural stability (Peixoto, 1962).

Peixoto's theorem. A C^r-vector field on a compact, two-dimensional manifold M^2 is structurally stable if and only if

(i) the number of fixed points and closed orbits is finite and each is hyperbolic;

(ii) there are no orbits connecting saddle points; and

(iii) the nonwandering set consists of fixed points and periodic orbits alone.

This theorem is the culmination of much previous work, especially by Poincaré (1899) and Andronov and Pontryagin (1937). It implies that typically a two-dimensional, structurally stable system will contain only sinks, saddles, sources, and repelling and attracting closed orbits in its invariant set, as shown in Figure 5.9a. Figure 5.9b shows some structurally unstable systems.

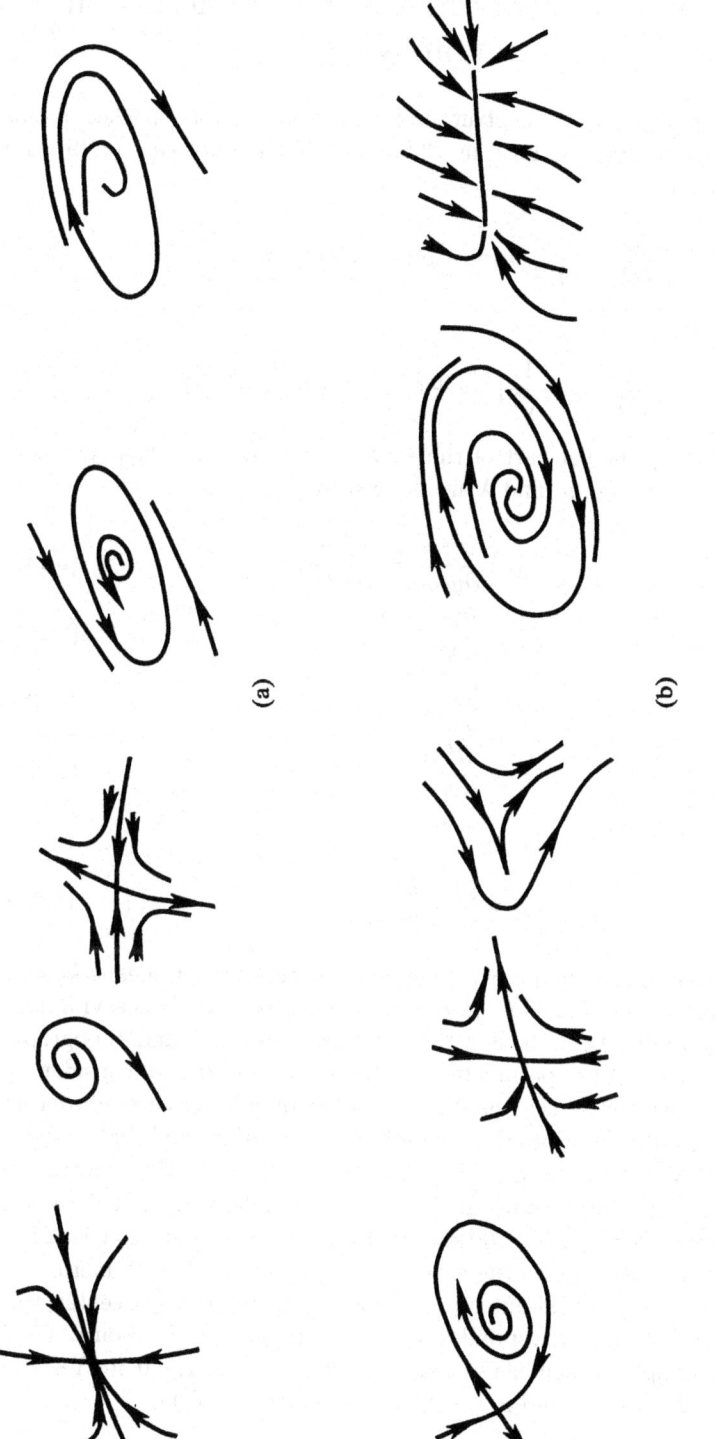

FIGURE 5.9. Phase plane diagrams for a two-dimensional dynamic system. (a) Structurally stable and (b) structurally unstable.

5.3 Bifurcation Properties of Wave Packets on Symmetric Topography

The equations governing the structural change in the wave packet in this chapter are the same as those in Chapter 3. We rewrite eqs. (3.49) and (3.50) as follows:

$$\frac{D_g m}{DT} = \frac{1}{K^2}\{k_0 m - \mu n - k_1(m^2 + n^2)n\} \tag{5.26}$$

and

$$\frac{D_g n}{DT} = -\frac{1}{K^2}\{\lambda m + k_0 n + k_2(m^2 + n^2)m\}, \tag{5.27}$$

where only the principal part of the effect of the basic flow has been considered and k_0, k_1, k_2, μ, and λ are defined by

$$k_0 = \frac{\partial^2 \eta_B}{\partial y \partial X} = \frac{\partial^2 \eta_B}{\partial x \partial Y}, \tag{5.28a}$$

$$k_1 = \frac{\partial V}{\partial X}, \tag{5.28b}$$

$$k_2 = \frac{\partial U}{\partial Y}, \tag{5.28c}$$

$$\mu = \frac{\partial^2 \eta_B}{\partial x \partial X}, \tag{5.28d}$$

and

$$\lambda = \delta_0 - \frac{\partial^2 \eta_B}{\partial y \partial Y}. \tag{5.28e}$$

In the following, for simplicity, the symmetric topography means $k_0 = 0$, and the symmetric basic flow means $k_1 = 0$ or $k_2 = 0$. The asymmetric topography means $k_0 \neq 0$. For the assumed basic flow and topography k_0, k_1, k_2, μ, and λ are parameters. In the present study, we consider the topography parameter, that is, λ or μ, as the mainly varying bifurcation parameter. Here λ describes the joint effect of the δ-effect and the curvature of topography in the east–west direction, and μ describes the curvature of topography in the north–south direction. For example, a westerly jet, which has a shear of about $2\,\mathrm{m/s}$, $100\,\mathrm{km}$ out from the jet center, that is, $2\,\mathrm{m/s}$ per $100\,\mathrm{km}$, corresponds to the case where k_2 is about 0.02. A meridional jet, which has a shear of about $1\,\mathrm{m/sec}$, $100\,\mathrm{km}$ out from the jet center, that is, $1\,\mathrm{m/s}$ per $100\,\mathrm{km}$, corresponds to the case in which k_1 is about 0.01. A convex topography, which has such a distribution as $\eta = H - 0.5(xX + yY)$, corresponds to $\lambda = 0.5$ and $\mu = -0.5$ on the earth's β-plane.

5.3.1 THE LARGEST SPATIAL SCALE EQUILBRIUM STATE

From eqs. (5.26) and (5.27), it is readily seen that the largest spatial scale state of a wave packet at the origin in the WKB phase space is always the equilibrium state of the system, that is,

$$m = 0, \qquad n = 0. \tag{5.29}$$

This state is called the *largest spatial scale state*. This state is not a valid state for the method used here. However, we can symbolically use this state as if it were a valid state without losing its dynamic property, as justified in Appendix A.

The linearized equations near this state will be

$$\frac{D_g m}{DT} = \frac{1}{F}(k_0 m - \mu n) \tag{5.30}$$

and

$$\frac{D_g n}{DT} = -\frac{1}{F}(\lambda m + k_0 n). \tag{5.31}$$

The stability factor p is readily found as

$$p^2 = \frac{k_0^2 + \mu\lambda}{F^2}. \tag{5.32}$$

From the results, we could conclude that near this state $m = n = 0$ is a saddle point in the WKB phase space when $\mu\lambda > -k_0$, which corresponds to a unstable manifold and a stable manifold through this state, as shown in Figure 5.10a. In this case, the largest scale state is unstable. On the other hand, if $\mu\lambda < -k_0^2$, then state $m = n = 0$ will be a center in the WKB phase space, as shown in Figure 5.10b. The manifold near $m = n = 0$ will be the closed curve, which indicates that the wave packet's structural vacillation (oscillation), found in Chapter 4, exists. Therefore

$$\lambda_c = -\frac{k_0^2}{\mu} \quad \text{or} \quad \mu_c = -\frac{k_0^2}{\lambda}, \tag{5.33}$$

is a critical value for the onset of the wave packet vacillations near the largest scale wave packet. In the following, the strong concave topography or concave topography means that $\lambda < 0$. The weak concave or convex topography means that $\lambda > 0$. As shown in Chapter 4 or by Yang (1988a), the δ-effect only modifies the east–west-oriented topography. Thus, in the following, the concave topography is referred to the case $\lambda < 0$ or $\mu > 0$, whereas the convex topography is referred to the case $\lambda > 0$ or $\mu < 0$.

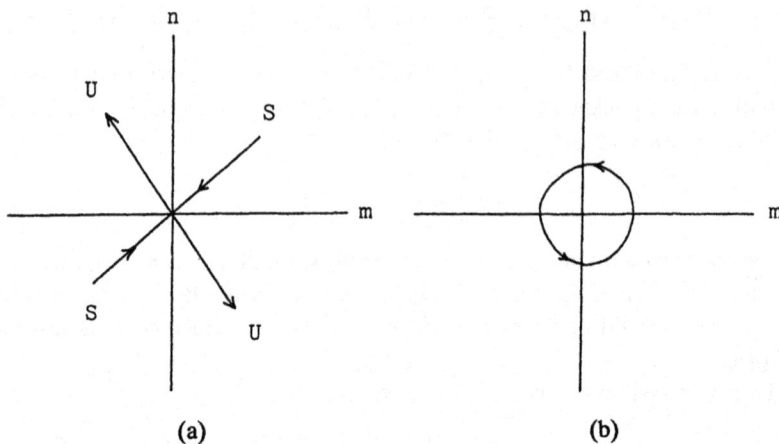

FIGURE 5.10. The manifolds near the largest spatial scale equilibrium state. (a) When the largest spatial scale is unstable and (b) when the largest spatial scale has neutral stability. Here U indicates the unstable manifold; S, the stable manifold.

5.3.2 ZONAL BASIC FLOW ON SYMMETRIC TOPOGRAPHY

In this case, the governing equations will become

$$\frac{D_g m}{DT} = -\frac{\mu}{K^2} n \tag{5.34}$$

and

$$\frac{D_g n}{DT} = -\frac{1}{K^2}\{\lambda m + k_2(m^2 + n^2)m\}. \tag{5.35}$$

The system has two kinds of equilibrium states, that is,

$$m = 0, \qquad n = 0, \tag{5.36}$$

and

$$m^2 = -\frac{\lambda}{k_2}, \qquad n = 0. \tag{5.37}$$

Therefore, there are two kinds of possible equilibria of the wave packet for the zonal basic flow over the symmetric topography. One state corresponds to the largest spatial scale state as discussed above. The other is one that has the largest latitudinal scale and a finite longitudinal scale, called the *purely longitudinal scale state*, when $\lambda k_2 < 0$. Therefore the latter only exists on a certain range of the topography parameters, depending on the position of the wave packet relative to the zonal basic flow. It should be pointed out that, in the WKB phase space, there are actually three equilibrium states, since there are two purely longitudinal scale states. However, in the following we will not distinguish between the two purely longitudinal scale states, since they share the same dynamic characteristics. In the case in which the wave packet is located on such a westerly jet and such a convex topography, as mentioned above, the equilibrium states (5.37) have the spatial scale with wave number 5. This spatial scale is a typical scale for the synoptic system in the atmosphere.

The linearized equations near the purely longitudinal scale state will become

$$\frac{D_g m}{DT} = -\frac{\mu}{K_M^2} n \qquad (5.38)$$

and

$$\frac{D_g n}{DT} = \frac{2\lambda}{K_M^2} m, \qquad (5.39)$$

where

$$K_M^2 = -\frac{\lambda}{k_2} + F, \qquad (5.40)$$

m and n are perturbations about the purely longitudinal scale states, and the primes in m and n have been omitted. It is readily found that the stability factor is

$$p^2 = -\frac{2\lambda\mu}{K_M^4}. \qquad (5.41)$$

Therefore, the bifurcated states are unstable when $\mu\lambda$ is less than zero, and they have neutral stabilities when $\mu\lambda$ is greater than zero.

The results show that on the left-hand side of a westerly jet the purely longitudinal scale states have neutral stabilities on concave topography. In this case, the bifurcation diagram is shown in Figure 5.11a, for unstable equilibrium states and neutral or stable equilibrium states. In this figure and the following ones, we present the local wave number in the X-direction to represent the longitudinal spatial scale, the local wave number in the Y-direction to represent the latitudinal spatial scale, and the total local wave number, defined by $(m^2 + n^2)^{\frac{1}{2}}$, to present the whole spatial scale because of the special relation between the local wave number and the spatial scale. The results demonstrate that the longitudinal spatial scale and the whole spatial scale of a wave packet on the left-hand side of a westerly jet has only one equilibrium state on the strong concave topography, that is, $\lambda < 0$. As soon as the topography parameter increases (or the strength of concave topography decreases) to a critical value, the largest scale state loses its stability, becoming unstable immediately, and the other two equilibrium states, that is, the purely longitudinal scale states that have neutral stabilities, appear. Now there are three equilibria, one unstable and two neutral, as shown in Figure 5.11a. Therefore, the bifurcation is a supercritical bifurcation in this case. However, the latitudinal scale state always remains the

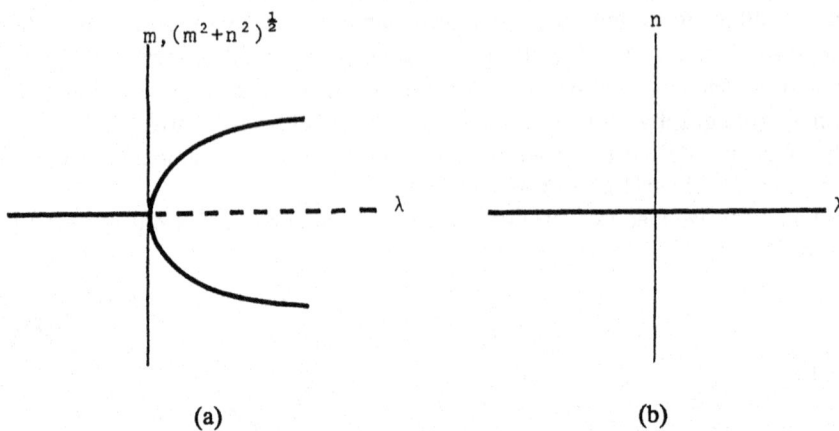

$$(a) \qquad\qquad\qquad\qquad (b)$$

FIGURE 5.11. A supercritical bifurcation diagram on symmetric concave topography when the wave packet is located on the left-hand side of a westerly jet. (a) For the longitudinal spatial scale equilibrium states and the whole spatial scale states, indicated by the local wave number in the X-direction and the local total wave number, respectively. (b) For the latitudinal spatial scale states, indicated by the local wave number in the Y-direction. Key: - - -, unstable equilibrium states; —, neutral equilibrium states.

same with no bifurcation for it, as shown in Figure 5.11b. Figure 5.11a corresponds to the supercritical bifurcation for the whole spatial scale states and the longitudinal scale states.

On the right-hand side of a westerly jet, the case is different. Figure 5.12 shows, on concave-topography the bifurcation diagram of a wave packet on the right-hand side of a westerly jet, where again there are equilibrium states and neutral or stable equilibrium states. It can be seen from Figure 5.12a that when the topography parameter is less than its critical value there are three equilibrium states. The largest spatial scale state is neutral and the purely longitudinal scale states are unstable. When the parameter crosses the critical value, the largest spatial scale loses its stability, and the purely longitudinal scale states disappear. Thus, the bifurcation in such a case is a subcritical bifurcation. The whole spatial scale states and the longitudinal spatial scale states both exhibit the subcritical bifurcation property as shown in Figure 5.12a. The latitudinal scale state is shown in Figure 5.12b. Now the latitudinal scale state also loses its stability when the parameter crosses the critical value, in contrast to the above. But we should point out that this bifurcation can only happen on weak concave topography. The situation, however, will occur as soon as the condition λ crosses zero is satisfied.

On convex topography, that is, $\lambda < 0$, we could draw conclusions, summarized in Figure 5.13 and Figure 5.14, about unstable and neutral equilibrium states. However, the present situation is different from the previous

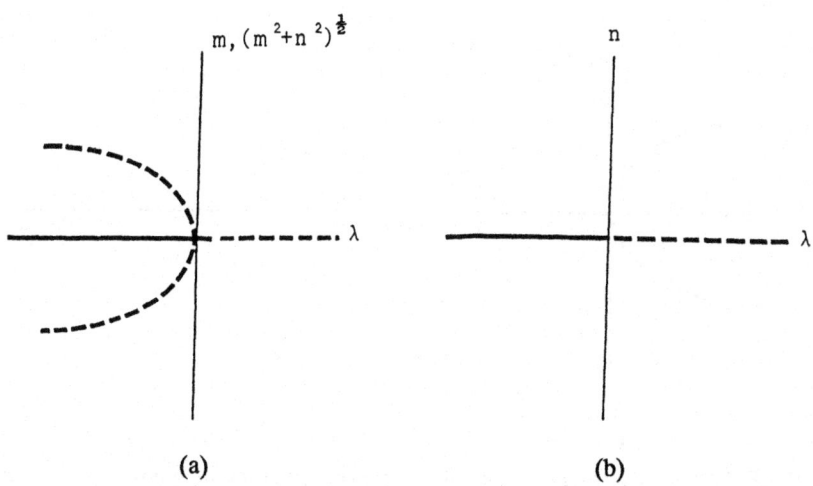

FIGURE 5.12. A subcritical bifurcation diagram on symmetric concave topography when the wave packet is located on the right-hand side of a westerly jet. (a) For the longitudinal spatial scale states and the whole spatial scale states and (b) for the latitudinal spatial scale states. Key: - - -, unstable equilibrium states; —, neutral equilibrium states.

one. Figure 5.13a is the bifurcation diagram for the whole spatial scale and the longitudinal spatial scale of a Rossby wave packet, which is located on the left-hand side of a westerly jet. When the topography parameter is less than its critical value, there exist three equilibrium states. One is the largest spatial scale state and the other two are the purely longitudinal scale states. In contrast to Figure 5.11, the bifurcated equilibrium states occur on the left side of the critical value, though the largest spatial scale state is still unstable and the bifurcated states have neutral stabilities.

As soon as the parameter crosses the critical value, the bifurcated states disappear and the largest scale state becomes a neutrally stable equilibrium state. The bifurcated states are neutrally stable when the parameter is less than its critical value. But compared with Figure 5.11a, Figure 5.13a is obviously a reverse Figure 5.11a. This is called *reverse bifurcation*. It is a *reverse supercritical bifurcation*. Figure 5.13b corresponds to the latitudinal scale state. The reverse supercritical bifurcation could not happen on real convex topography, since λ is always greater than zero. But the reverse supercritical bifurcation will happen as soon as the condition $\mu < 0$ and λ does cross zero. The condition could be satisfied on the certain saddle topography, in fact.

On the right-hand side of a westerly jet, however, the bifurcation property will be different. Figure 5.14 shows this situation. When the topography parameter is less than the critical value, there exists only one equilibrium state, that is, the largest scale state. This state is unstable. When the topog-

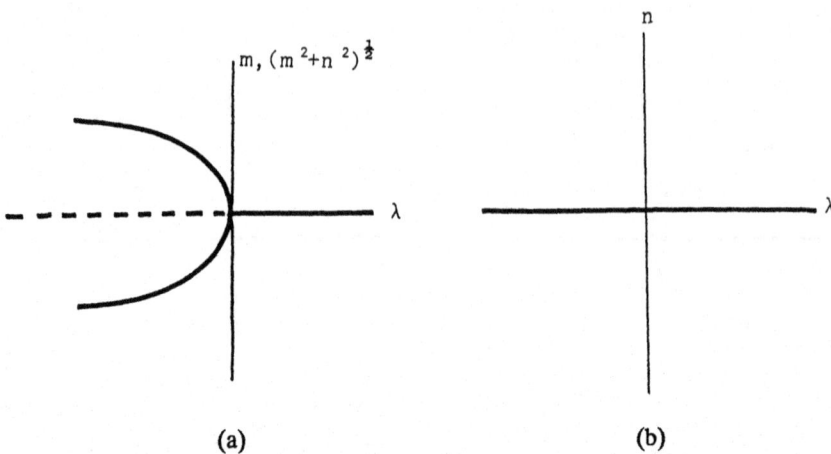

(a)   (b)

FIGURE 5.13. A reverse supercritical bifurcation diagram on symmetric convex topography when the wave packet is located on the left-hand side of a westerly jet. (a) For the longitudinal spatial scale states and the whole spatial scale states and (b) for the latitudinal spatial scale state. Key: - - -, unstable equilibrium states; —, neutral equilibrium states.

raphy parameter crosses the critical value, the largest scale state become neutrally stable and two additional bifurcated states occur. The bifurcated states are always unstable. However, compared with Figure 5.12, the diagram is a reverse one. Thus, from the results one could find that on the right-hand side of a westerly jet, the whole spatial scale state and longitudinal scale state of a Rossby wave packet will appear as a *reverse subcritical bifurcation*, as shown in Figure 5.14a. Figure 5.14b shows the latitudinal spatial scale state.

From the stability analysis, it can also be shown that the supercritically bifurcated states are time periodic, with the frequency ω_M, which is found to be

$$\omega_M = \frac{\sqrt{2|\mu\lambda|}}{K_M^2}. \tag{5.42}$$

The largest spatial scale state or, precisely, the state near the largest spatial scale state, is also time periodic, with frequency ω_0, which is easily shown to be

$$\omega_0 = \frac{\sqrt{2|\mu\lambda|}}{F}. \tag{5.43}$$

From the bifurcation diagrams, one can see that the wave packet will bifurcate from one state with frequency ω_0 into another state with frequency ω_M as λ crosses zero, as shown in Figure 5.11 and Figure 5.13. The time periodic states correspond to the wave packet vacillations (oscillations) found in Chapter 4, which are characterized by periodic changes in both the tilt and the spatial scales of a wave packet. Using the above orders of magnitude

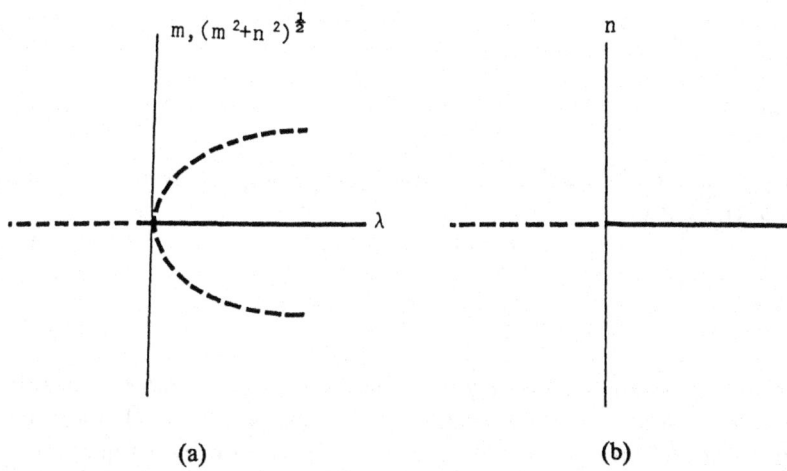

(a)   (b)

FIGURE 5.14. A reverse subcritical bifurcation diagram on symmetric convex topography when the wave packet is located on the right-hand side of a westerly jet. (a) For the longitudinal spatial scale states and the whole spatial scale states and (b) for the latitudinal spatial scale state. Key: - - -, unstable equilibrium states; —, neutral equilibrium states.

for the parameters, we find that the purely longitudinal scale states have a vacillation with a period of about 20 to 30 days in the dimensional time scale, whereas the largest spatial scale state has a vacillation with a period of about a few days in dimensional time scale. The bifurcations discussed here and in the following have meaningful implications in the evolution of a Rossby wave packet, especially in its structural changes. From the topological point of view, one could clearly see them in the WKB phase space, since at the bifurcation point the system is always structurally unstable. We discuss this in Section 5.5.

In fact, the complete WKB integral could be derived as follows, in the WKB phase space:

$$m^2 + n^2 + \frac{\lambda + \mu}{k_2} = Ce^{\frac{k_2}{\mu}m^2}, \tag{5.44}$$

where C is an integral constant to be determined by the initial condition.

5.3.3 MERIDIONAL BASIC FLOW ON SYMMETRIC TOPOGRAPHY

If we consider that the basic flow has only a meridional flow, which is easily observed near the north–south-oriented coasts in the ocean, and symmetric topography, the governing systems will become

$$\frac{D_g m}{DT} = -\frac{1}{K^2}\{\mu n + k_1(m^2 + n^2)n\} \tag{5.45}$$

and

$$\frac{D_g n}{DT} = -\frac{\lambda}{K^2} m. \tag{5.46}$$

Taking μ as a main varying bifurcation parameter, the system has two kinds of equilibria:

$$m = 0, \qquad n = 0; \tag{5.47}$$

and

$$m = 0, \qquad n^2 = -\frac{\mu}{k_1}. \tag{5.48}$$

The results show that the *purely latitudinal spatial scale states* exist only for a certain range of the parameter μ. For the case in which the wave packet is located on such a meridional jet and such a convex topography, as considered above, the equilibrium states (5.48) have the spatial scale with wave number 7. This is a typical spatial scale for the synoptic system in the atmosphere. The system now is very similar to the previous one, though the physical meaning is different. Compared with the case of zonal basic flow, now the purely latitudinal scale state replaces the purely longitudinal scale state and the meridional flow instead of the zonal flow.

The results are shown in Figure 5.15, with unstable equilibrium states and neutral or stable equilibrium states. Figure 5.15a and Figure 5.15b are the bifurcation diagrams for weak concave or convex topography, that is, $\lambda > 0$. This condition is easily satisfied on the earth's δ-surface. On the earth's β-plane, the condition that $\lambda > 0$ means that the topography has to be convex, that is, $(\partial^2 \eta_B / \partial y \partial Y) < 0$. Figure 5.15c and Figure 5.15d correspond to the bifurcation diagrams for strong concave topography, that is, $\lambda < 0$. Figure 5.15a shows the bifurcation diagrams for the whole spatial scale and latitudinal scale states when the wave packet is located on the left-hand side of a meridional flow and on weak concave or convex topography. In this case, the bifurcation is supercritical, and the bifurcation state is time periodic with frequency ω_N, which is readily found to be

$$\omega_N = \frac{\sqrt{2|\mu\lambda|}}{K_N^2}, \tag{5.49}$$

where K_N^2 is defined by

$$K_N^2 = -\frac{\mu}{k_1} + F. \tag{5.50}$$

Using the above orders of magnitude for the parameters, the wave packet structural vacillation has a period of order of $O(10)$ days in dimensional time scale. Figure 5.15b corresponds to the case in which the wave packet is located on the right-hand side of a meridional flow. The result shows that this bifurcation is subcritical. Figure 5.15c shows the bifurcation diagram when the wave packet is located on strong concave topography and on the

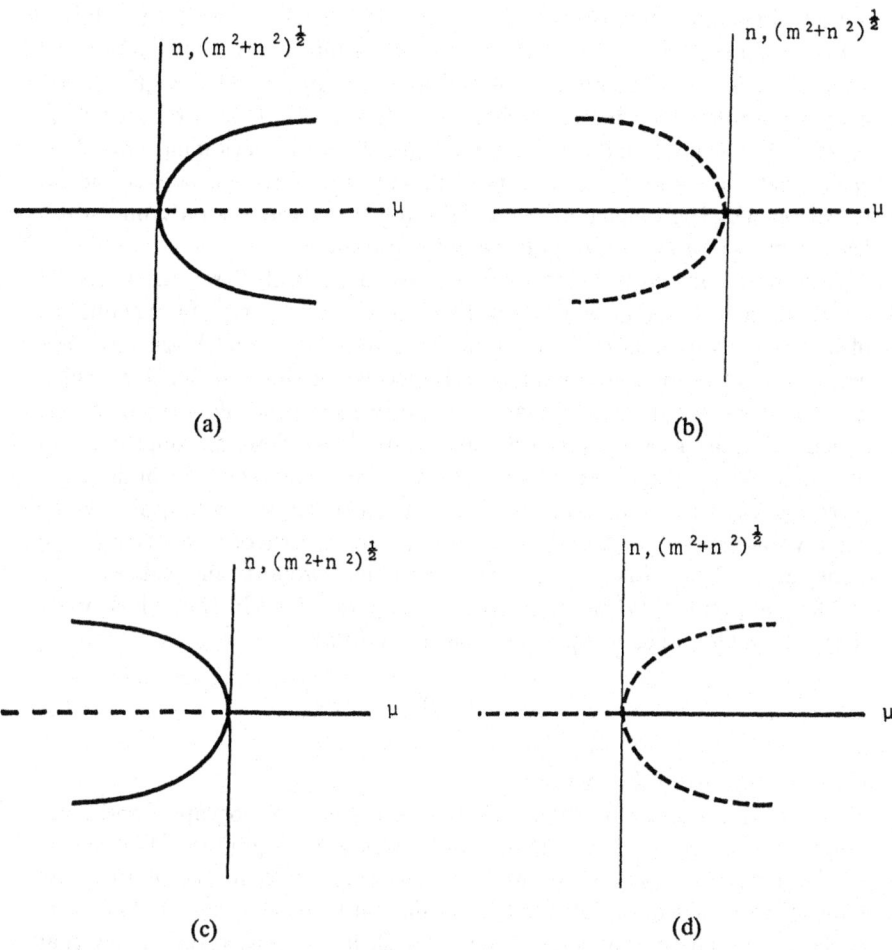

FIGURE 5.15. Bifurcation diagrams on symmetric topography for the whole spatial scale states and latitudinal spatial scale states in a meridional basic flow. On convex topography, when the wave packet is located on the right-hand side (a) and on the left-hand side (b) of a meridional flow; on the strong concave topography when the wave packet is located on the left-hand side (c) and on the right-hand side (d) of a meridional flow. Key: - - -, unstable equilibrum states; —, neutral equilibrium states.

left-hand side of a meridional flow. The diagram indicates that the bifurcation in such a case is a reverse supercritical bifurcation. The bifurcated scale state is still time periodic with frequency ω_N. Finally, Figure 5.15d gives the bifurcation diagram when the wave packet is located on strong concave topography and on the right-hand side of a meridional flow. It is easily seen that, in such a case, the bifurcation is a reverse subcritical bifurcation. For the longitudinal spatial scale state, there is only one state, that is, $m = 0$, and we will not present its results.

A comparison with the results in a zonal basic flow reveals that the effect of a meridional basic flow on the bifurcation property of the evolution of a Rossby wave packet is different from that of a zonal basic flow. For example, on the concave topography, the evolution of a wave packet, which is located on the left-hand side of a symmetric basic flow exhibits the supercritical bifurcation property in a zonal basic flow, although instead of a meridional basic flow it exhibits the reverse supercritical bifurcation property. However, on convex topography, the evolution will appear to be the reverse subcritical property on the right-hand side of a westerly jet; in a meridional basic flow, it shows the subcritical property instead.

The complete WKB integral could also be obtained in the WKB phase space, as we found in the purely zonal basic flow:

$$m^2 + n^2 + \frac{\lambda + \mu}{k_1} = Ce^{\frac{k_1}{\lambda}n^2}, \tag{5.51}$$

where C is a integral constant.

The results also demonstrate that on symmetric topography and symmetric basic flow, there are three kinds of possible equilibria. The first is the largest spatial scale state, with $m = n = 0$. The second is the two pure scale states: the purely longitudinal scale states, $m \neq 0$, $n = 0$. The third is the purely latitudinal scale states, $m = 0$, $n \neq 0$. The results show that on symmetric topography it is impossible for the mixed scale equilibrium states to exist, that is, $mn \neq 0$, in a symmetric basic flow. However, the asymmetry of topography will allow the mixed scale states to exist, as we show in the following.

5.4   Bifurcation Properties of Wave Packets on Asymmetric Topography

5.4.1   ZONAL BASIC FLOW

When only zonal basic flow is involved, the governing equations with asymmetric topography become

$$\frac{D_g m}{DT} = \frac{1}{K^2}\{k_0 m - \mu n\} \tag{5.52}$$

and

$$\frac{D_g n}{DT} = -\frac{1}{K^2}\{\lambda m + k_0 n + k_2(m^2 + n^2)m\}. \tag{5.53}$$

In this case, the two kinds of equilibria could be easily found as follows:

$$m = 0, \qquad n = 0, \tag{5.54}$$

$$m^2 = M^2 = -\frac{\mu(k_0^2 + \lambda\mu)}{k_2(k_0^2 + \mu)}, \tag{5.55a}$$

and

$$n = N = \frac{k_0}{\mu}M. \tag{5.55b}$$

In contrast to the previous case, one finds that for the present system, the asymmetry of the topography makes it possible for the mixed scale states to exist. However, one also finds that now it is impossible for the purely longitudinal scale state or the purely latitudinal scale state to occur. The expressions (5.55a) and (5.55b) state that the mixed scale states exist only when the condition that the right-hand side of (5.55a) is greater than zero is satisfied. They are

$$\lambda < \lambda_c, \tag{5.56}$$

when $k_2 > 0$, and

$$\lambda > \lambda_c, \tag{5.57}$$

when $k_2 < 0$, where λ_c is defined by (5.33).

We still take the topography parameter λ as the mainly varying bifurcation parameter. The stability for the largest scale state is the same as before. The stability factor is shown in (5.32). For mixed scale states, however, we obtain the following linearized equation to determine their stabilities:

$$\frac{D_g m}{DT} = \frac{1}{K_{MN}^2}\{k_0 m - \mu n\} \tag{5.58}$$

and

$$\frac{D_g n}{DT} = -\frac{1}{K_{MN}^2}\{\lambda m + k_0 n + k_2[(3M^2 + N^2)m + 2MNn]\}, \tag{5.59}$$

where m and n are perturbations about the mixed scale states; the primes in m and n have been omitted; M and N are the equilibrium values for the mixed scale states defined by (5.55a) and (5.55b) and K_{MN}^2 is defined by

$$K_{MN}^2 = M^2 + N^2 + F \tag{5.60a}$$

or

$$K_{MN}^2 = -\frac{k_0^2 + \lambda\mu}{\mu k_2} + F. \tag{5.60b}$$

The characteristic equation governing the stability of the mixed scale state is found to be

$$p^2 + 2bp + c = 0, \tag{5.61}$$

where

$$b = \frac{1}{K_{MN}^2} MNk_2 \tag{5.62a}$$

and

$$c = -\frac{1}{K_{MN}^2}\{k_0^2 + 2MNk_2k_0 + \mu\lambda + \mu k_2(3M^2 + N^2)\}. \tag{5.62b}$$

Using the criterion of relations between the roots and coefficients in the quadratic algebraic equation, we could conclude that when the topography parameter is greater than its critical value, which is defined by

$$\lambda_{MN} = \lambda_c = -\frac{k_0^2}{\mu}, \tag{5.63}$$

the mixed scale states are stable and they are unstable when the topography parameter is less than the critical value on concave topography, that is, $\mu > 0$. On convex topography, that is, $\mu < 0$, the mixed scale states are unstable when the topography parameter is greater than the critical value, whereas they are stable when the topography parameter is less than its critical value.

According to the above analysis, one obtains different bifurcation diagrams in different cases, as shown in Figure 5.16 and Figure 5.17, with unstable and stable or neutral equilibrium states. Compared with the previous results, the main difference is that on asymmetric topography, it is possible for mixed equilibrium states to exist. Figure 5.16 illustrates the bifurcation in such a case, in which the wave packet on concave topography is located on the right-hand side of a westerly jet. When the topography parameter is less than the critical value, there are two kinds of equilibrium states, that is, the largest scale state and two mixed scale states. According to the above stability analysis, we could conclude that the mixed scale states are unstable but that the largest scale state has neutral stability.

As soon as the parameter crosses the critical value, the largest scale state will lose its stability, becoming unstable, and the mixed scale states disappear immediately. Figure 5.16a corresponds to the bifurcation diagram for the longitudinal scale states. Figure 5.16b and Figure 5.16c are for the latitudinal scale states and the whole spatial scale states, respectively. The diagrams show that it is a subcritical bifurcation. Now the bifurcation critical value is not zero but rather λ_{MN}. The critical value could only be reached on strong concave topography.

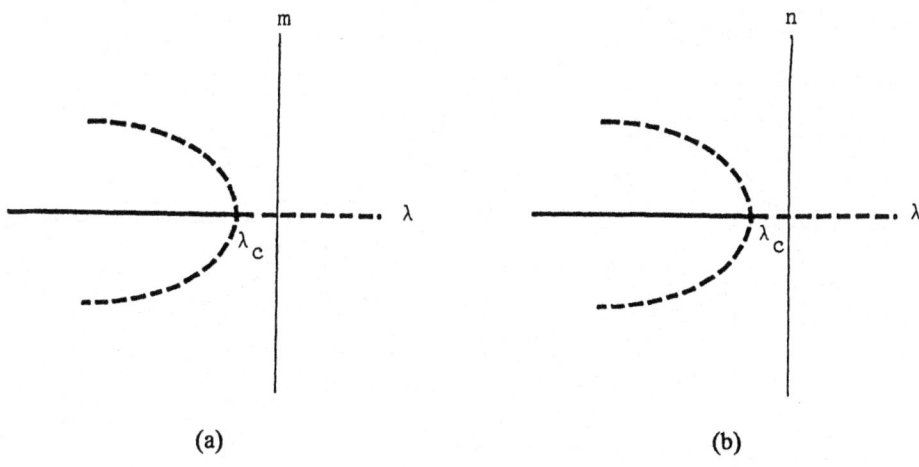

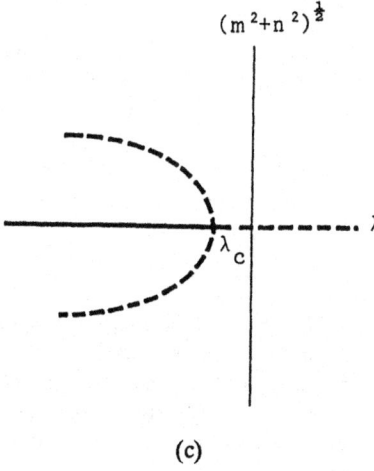

FIGURE 5.16. Bifurcation diagrams on asymmetric concave topography when the wave packet is located on the right-hand side of a westerly jet. (a) For the longitudinal spatial scale states, (b) for the latitudinal spatial scale states, and (c) for the whole spatial scale states. Key: - - -, unstable equilibrium states; —, stable or neutral equilibrium states.

From the figures, we find that the bifurcation diagrams for the longitudinal scale state, the latitudinal scale state, and the whole scale state all exhibit subcritical bifurcation behavior, whereas on symmetric topography, one of the scale states does not have this behavior, as discussed before. In the following, we present the bifurcation diagram for the whole spatial scale state, since the other two scale states behave similarly. Figure 5.17 shows the other three cases. Figure 5.17a corresponds to the case in which the wave packet on the concave topography is located on the left-hand side of a westerly jet. From the figure, it is found that when the topography parameter is less than its critical value, there is only one equilibrium state, with neutral stability.

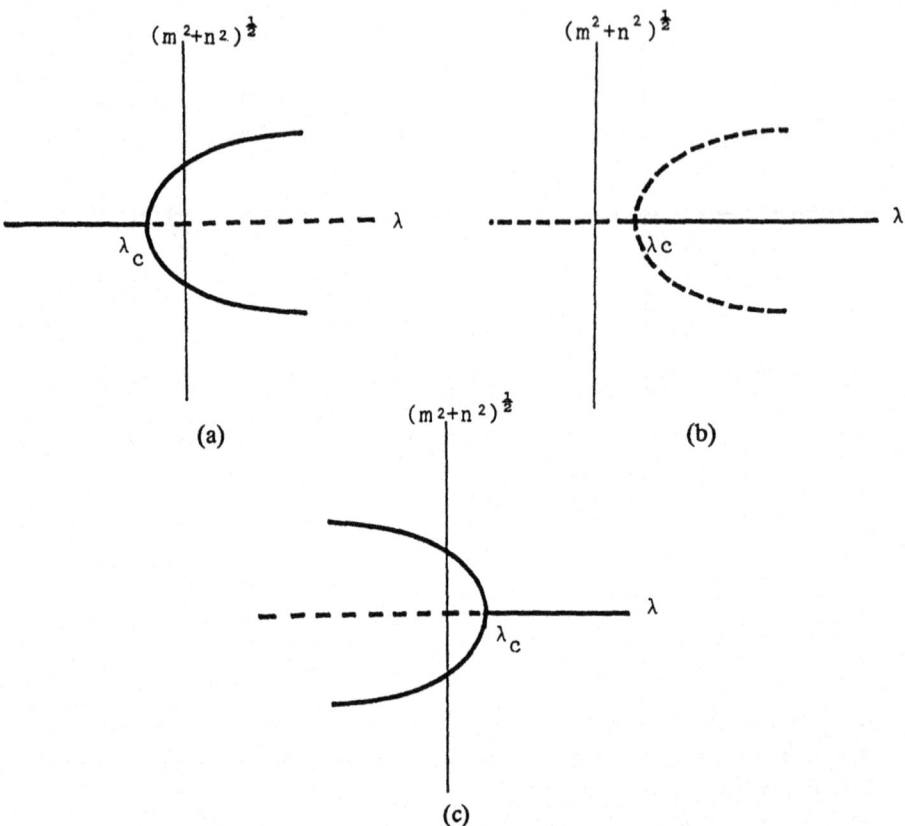

FIGURE 5.17. Bifurcation diagrams for the whole spatial scale states on asymmetric topography. (a) The wave packet on the concave topography is located on the left-hand side of a westerly jet; (b) the wave packet on the convex topography is located on the right-hand side of a westerly jet; and (c) on the left-hand side of a westerly jet. Key: - - -, unstable equilibrium states; —, neutral equilibrium states.

When the parameter crosses the critical value, the largest scale state becomes unstable and there appear two additional mixed scale states. The mixed scale states are stable. Thus, the bifurcation diagram shows that in such a case the bifurcation possess a supercritical property. Figure 5.17b and Figure 5.17c are the cases on convex topography. Figure 5.17b presents the result when the wave packet is located on the right-hand side of a westerly jet, and Figure 5.17c is the bifurcation diagram when the wave packet is located on the left-hand side of a westerly jet. From the bifurcation diagrams, one finds that, on convex topography, the bifurcations are the reverse ones. On the right-hand side and on the left-hand side of a westerly, the bifurcations possess reverse subcritical and supercritical properties, respectively.

5.4.2 MERIDIONAL BASIC FLOW

For a meridional basic flow, we have the following system:

$$\frac{D_g m}{DT} = \frac{1}{K^2}\{k_0 m - \mu n - k_1(m^2 + n^2)n\} \tag{5.64}$$

and

$$\frac{D_g n}{DT} = -\frac{1}{K^2}\{\lambda m + k_0 n\}. \tag{5.65}$$

It is readily found that there are two kinds of possible equilibrium states, that is,

$$m = 0, \qquad n = 0, \tag{5.66}$$

$$m = M = -\frac{k_0}{\lambda}, \tag{5.67a}$$

and

$$n^2 = N^2 = -\frac{\lambda}{k_1}\frac{k_0^2 + \mu\lambda}{k_0^2 + \lambda^2}. \tag{5.67b}$$

Equation (5.66) is the largest spatial scale state, whereas eqs. (5.67a,b) correspond to the two mixed scale equilibrium states.

Taking μ as a main changing bifurcation parameter, we obtain bifurcation diagrams in various situations. As on symmetric topography, the effect of a meridional basic flow on the bifurcation property of the evolution of a wave packet is different from that of a zonal basic flow. The difference between the two kinds of symmetric basic flows at present is almost the same as on symmetric topography. Except for these differences, however, the bifurcation diagrams will closely resemble the case of zonal basic flow discussed above. Therefore we will not present the diagrams here. It can be shown that the bifurcation critical value is μ_c, which is defined by (5.33). The results demonstrate that the bifurcation is supercritical (subcritical) when the wave packet is located on the left- (right-) hand side of a meridional flow with weak concave or convex topography, that is, $\lambda > 0$. The bifurcation diagrams show that the bifurcation is the reverse supercritical (subcritical) bifurcation when the wave packet is located on the left- (right-) hand side of a meridional flow with convex topography.

The above results demonstrate that on the same type of topography (either the symmetric topography or the asymmetric topography), the wave packet bifurcation on one side of a basic flow differs from that on the other side of the basic flow. For example, on convex topography, the bifurcation is a reverse supercritical bifurcation on the left-hand side of a zonal basic flow, whereas on the right-hand side of the basic flow, the bifurcation is a reverse subcritical one. Therefore, the topological structure of the evolution of a

Rossby wave packet in the WKB phase space will be completely different from one side of a basic flow to the other side. The system governing the evolution of a wave packet is thus structurally unstable at the center of a basic flow. Taking the strength of the shear of a basic flow as a varying bifurcation parameter, we can see that bifurcation also occurs when the wave packet passes the center of a basic flow. From a physical point of view, however, the results suggest that a wave packet would evolve completely differently from one side of a basic flow to the other side. These predictions are often confirmed in real geophysical flows.

5.5 Trajectories in the WKB Phase Space

In order to better understand the bifurcation properties found above, we now use the WKB phase space to see what it will look like in such a phase space; this has been used in Chapter 4. The bifurcation properties actually describe the asymptotic characteristics of a system. In the present case, they are long time properties of the evolution of a wave packet. The use of the WKB phase space helps us to see how such a wave packet evolves under the various conditions discussed above, from the more physical point of view.

From (5.26) and (5.27), we find the divergence of the WKB phase space:

$$\frac{\partial}{\partial m}\left(\frac{D_g m}{DT}\right) + \frac{\partial}{\partial n}\left(\frac{D_g n}{DT}\right) = \frac{2mn}{K^2}\{(\lambda + \mu) - F(k_1 + k_2)\}, \qquad (5.68)$$

which states that the divergence changes its sign in the phase space. Therefore, according to the Poincaré–Bendixson theorem and the Bendixson criterion (Hirsh and Smale, 1974; Andronov et al., 1966; Guckenheimer and Holmes, 1983), it is possible for the closed orbit, that is, the wave packet vacillation, to exist. We know, however, that for two-dimensional systems such as (5.26) and (5.27), all possible nonwandering sets will always fall into the following three classes (e.g., Andronov et al., 1966):

(i) fixed points;

(ii) closed orbits; and

(iii) the unions of fixed points and the trajectories connecting them.

When the unions of fixed points and the trajectories connecting them connect distinct points, they are called *heteroclinic orbits;* when they connect a point to itself they are called *homoclinic orbits* (e.g., Guckenheimer and Holmes, 1983; Sparrow, 1982), which are often separatrices of the closed orbits.

Now let us look at the topological structural changes in the WKB phase space and their implications in the evolution of a wave packet as the bifurcation parameter crosses the bifurcation point. On concave topography,

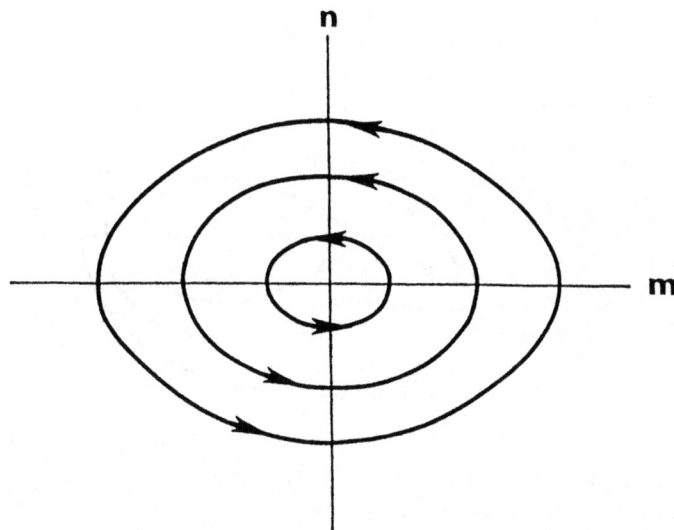

FIGURE 5.18. The WKB phase space. The wave packet on symmetric concave topography is located on the left-hand side of a westerly jet when the topography parameter is less than the critical value, where $\lambda = -1$, $\mu = 1$, and $k_2 = -1$ have been taken. The arrows indicate the flow direction of the evolution of a Rossby wave packet.

for example, as shown in Figure 5.11, the evolution of a Rossby wave has a supercritical bifurcation on the left-hand side of a westerly jet as the topography parameter crosses the critical value. When the parameter is less than its critical value, there is only one equilibrium state in the WKB phase space. This equilibrium state has neutral stability. Therefore, in the WKB phase space, this equilibrium state corresponds to a center at the origin point as shown in Figure 5.10b. Since there is no other fixed point on the entire phase space, except perhaps at infinity, according to the two-dimensional differential equation theory, there are closed orbits at least on some subset of the phase space near the origin point.

Figure 5.18 is a typical example. Here we take $\lambda = -1$, $\mu = 1$, and $k_2 = -1$. From the figure, one finds that on the closed orbit domain, the wave packet will evolve a wave packet structural vacillation (oscillation), which has been shown to be very similar to the wave vacillations found in geophysical fluids, as seen in Chapter 4. Both the tilt and the spatial scales of a wave packet, generally, exhibit time periodic changes simultaneously. The tilt of a wave packet in this case can tilt westward, for example, north–west to south–east, then tilt eastward, north–east to south–west, and later westward.

As soon as the topography parameter crosses the critical value, the largest spatial scale state loses its stability and become unstable, while

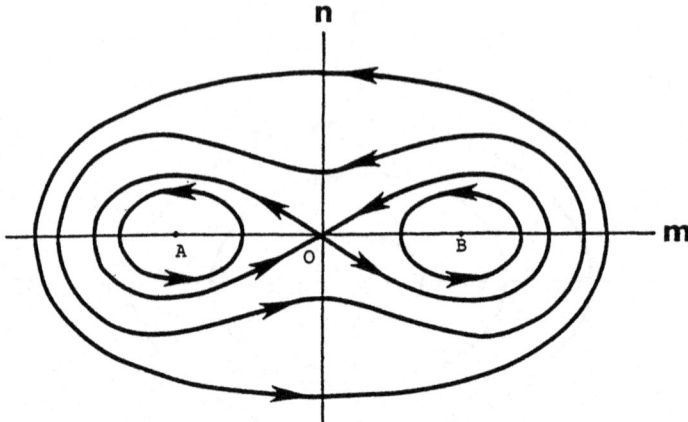

FIGURE 5.19. The same as in Figure 5.18, except that the topography parameter is greater than its critical value, taking $\lambda = 1$.

there appear two additional purely longitudinal scale states. The bifurcated states have neutral stabilities. Therefore, the topological structure on the WKB phase space differs from the previous one. In the WKB phase space, according to the differential equation theory, the origin point now is a saddle point crossed by two unstable manifolds and two stable manifolds, as shown in Figure 5.10a. The bifurcated states in the WKB phase space now correspond to two centers along the n-axis. Figure 5.19 is an example of this case, where $\lambda = 1$, $\mu = 1$, and $k_2 = -1$ have been taken. We find that closed orbits still exist, which suggests that wave packet structural vacillations are occurring.

Moreover, from the results, we also find that there are different kinds of wave packet structural vacillations. When the wave packet is initiated near the center A($-1,0$) or B(1.0), the longitudinal spatial scale is always finite. Therefore, the wave packet's structural vacillation is restricted to a range of longitudinal scale and the tilt vacillates in a certain manner. For example, the tilt of the wave packet is not parallel to the direction of north. There exists a special trajectory passing through the origin the WKB phase space. This trajectory separates one kind of the wave packet structural vacillation from other kinds. This special trajectory is the *homoclinic orbit* of the system. The results suggest that, in this case, the wave packet can change its characteristics from one kind of vacillation to another kind, as the parameter is varied.

On the right-hand side of a westerly jet, with concave topography, however, the story is different. The bifurcation diagram is shown in Figure 5.12. When the topography parameter is less than its critical value, there are three equilibrium states located at O($0,0$), A($-1,0$), and B($1,0$) in the WKB phase space, as shown in Figure 5.20. In this figure, we take $\lambda = -1$,

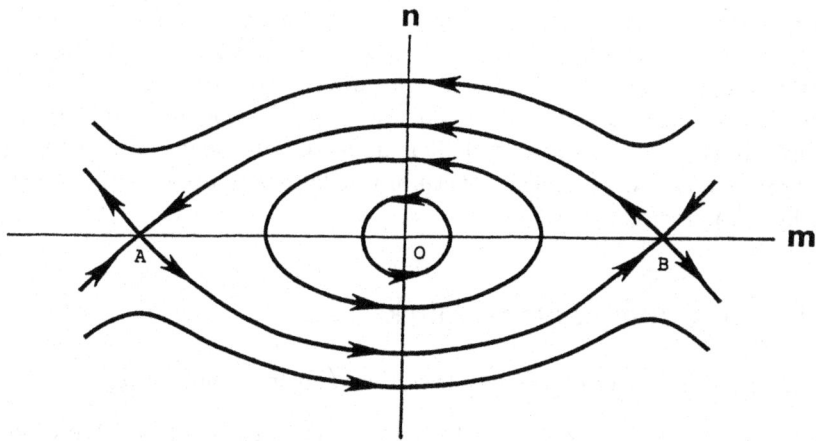

FIGURE 5.20. The WKB phase space. The wave packet on the symmetric concave topography is located on the right-hand side of a westerly jet, where $\lambda = -1$, $\mu = 1$, and $k_2 = 1$ have been taken. The arrows indicate the flow direction of the evolution of a wave packet.

$\mu = 1$, and $k_2 = 1$. Since the largest spatial scale state has neutral stability, the origin $O(0,0)$ is a center. The bifurcated states are unstable. Thus, $A(-1,0)$ and $B(1,0)$ are saddle points in the WKB phase space. Near the center, the wave packet shows structural vacillation. Special trajectories connect $A(-1,0)$ and $B(1,0)$. These orbits are the *heteroclinic orbits*, which are the separatrices from the wave packet's structural vacillations. The results suggest that, in this case, the wave packet can change its characteristics from one kind of wave packet structural vacillation to another kind of wave packet structural vacillation, as the topography parameter is varied.

When the topography parameter is greater than its critical value, the WKB phase space diagram is similar to Figure 5.18. We could also obtain the WKB phase space diagrams for the meridional basic flow, which are similar to those for a zonal basic flow. We have not presented these diagrams. For the meridional basic flow, however, the evolution of a wave packet has yet another kind of wave packet structural vacillation when the bifurcation is supercritical or reverse supercritical. This kind of vacillation is characterized by the vacillation being restricted to a range of latitudinal spatial scale instead of a longitudinal spatial scale in the zonal basic flow. The transition can only occur between this kind of wave packet structural vacillation and the first kind of wave packet structural vacillation characterized by its center at the origin in the WKB phase space, as the topography parameter is varied.

On asymmetric topography, however, for the supercritical and the reverse supercritical bifurcation cases, that is, on the left-hand side of a symmetric basic flow, one observes that the wave packet structure will asymptotically approach some patterns when the wave packet is initiated on some domain of the WKB phase space. Therefore, the wave packet asymptotically tilts westward or eastward. There are no wave packet structural vacillations. Thus, on asymmetric topography, it is impossible for transitions between different kinds of wave packet structural vacillations to occur as the topography parameter changes.

5.6   Summary and Remarks

From the above analysis, we can draw the following conclusions:

1. On both the symmetric and the asymmetric topographies, there can exist supercritical, subcritical, reverse supercritical, and reverse subcritical bifurcations in a symmetric basic flow, as the topography changes.

2. The effect of a zonal basic flow on the bifurcation property differs from that of a meridional basic flow.

3. On symmetric topography, three kinds of equilibrium states, that is, the largest spatial scale state, the two purely longitudinal spatial scale states, and the two purely latitudinal spatial scale states, can exist. On asymmetric topography, mixed spatial scale states occur, which suggests that on asymmetric topography the structure of a wave packet may asymptotically approach a certain pattern with the westward or eastward tilt when the mixed spatial scale states are stable.

4. The homoclinic orbits and heteroclinic orbits have been found in the WKB phase space; they are separatrices of the closed orbits. These results suggest that different kinds of wave packet vacillations exist.

5. The results also suggest that the evolution of a Rossby wave packet on one side of a basic flow can differ from that on the other side. These differences are often expected in real geophysical flows. At the center of a basic flow, the dynamic system governing the evolution of a wave packet is also structurally unstable, which suggests that the basic flow also plays a very important role in the bifurcation property.

The bifurcation properties of evolution of the wave packet considered above are also summarized in Table 5.1. The first column of the table lists the conditions of parameters, the second column, the equilibria. The third and fourth columns are types of bifurcation and simple bifurcation diagrams, respectively. One case is for zonal basic flow, the other case is for meridional basic flow (in parentheses).

TABLE 5.1. Summary of bifurcation properties.

Parameter conditions	Equilibria	Type of bifurcation	Diagrams	
			On symmetric topography	On asymmetric topography
$k_2 < 0; \mu > 0$ $(k_1 > 0; \lambda > 0)$	On symmetric topography $m = 0, n = 0$; $m^2 = \lambda/k_2$, $n = 0$.	Supercritical		
$k_2 > 0; \mu > 0$ $(k_1 < 0; \lambda > 0)$	$(m = 0, n = 0$; $m = 0$, $n^2 = -\mu/k_1.)$	Subcritical		
$k_2 < 0; \mu < 0$ $(k_1 > 0; \lambda < 0)$	On asymmetric topography $m = 0, n = 0$; $m = M, n = N.$	Reverse supercritical		
$k_2 > 0; \mu < 0$ $(k_1 < 0; \lambda < 0)$	$(m = 0, n - 0$; $m = M, n = N)$	Reverse subcritical		

There is a total of three kinds of possible wave packet structural vacillations. The first kind is characterized by the largest spatial scale state as its center in the WKB phase space. The second kind is restricted to a range of longitudinal spatial scale. The third kind is restricted to a range of latitudinal scale. Our results show that transitions can occur only between the first kind and the second kind or the third kind of wave packet structural vacillations, as the topography parameter is varied. The condition that prevents transitions between the second kind of vacillation and the third kind of vacillation is the symmetry of the basic flow. As shown in Chapter 4, the wave packet structural vacillation found here is similar to that observed in geophysical fluids (e.g., Hide, 1958; Fultz et al., 1959; Pfeffer and Chiang, 1967; Elberry, 1968; McGuirk and Reiter, 1976; Gruber, 1975; Webster and Keller, 1975).

Bifurcation is one of the most important properties in the dynamics of geophysical fluids. The structural change of disturbance systems is of great importance because it plays a special role in predictions. The results connect the structure change of a disturbance system to its bifurcation property in a simplified model, which has been shown to be capable of describing the evolution of a Rossby wave packet, and, especially, its structural changes. It is rather interesting to note that the present theory could deal with these changes very neatly. In our work, nevertheless, only primary bifurcations have been found. The results in both the numerical studies and the experiment (e.g., Weng et al., 1986; Hart, 1984) suggest that there a cascade of bifurcation exists. Obviously, further investigations in the bifurcation property of geophysical flows are needed. In Chapter 6, the asymmetry of the basic flow will be taken into account. The results show that two distinct bifurcations occur as the topography parameter is varied. The primary bifurcation is still pitchfork bifurcation. The secondary bifurcation is always transcritical. Transitions thus can occur among the three kinds of wave packet structural vacillations, as the topography parameter is varied.

Appendix A. Justification of the Largest Spatial Scale State

The WKB method is only valid in the limit that the wave packet length $L_p = O(L)$ is much smaller than the length scale of variations in the medium $L_m = O(a)$, where a is the average radius of the earth, and L and a are the same as defined in Section 5.2. Obviously, this precludes $m = n = 0$ as a valid center wave number for the wave packet. However, we can show that there exists a region in the WKB phase space, that is, m and n phase space, where m and n are sufficiently small so that the dynamic behavior near this region is the same as at $m = n = 0$, but still large compared to the inverse spatial scale of the background state, so the

WKB method is valid outside the region.

We define a small number as

$$\varepsilon = \frac{L_p}{L_m} = O\left(\frac{L}{a}\right) \ll 1. \tag{5.69}$$

Note that the small number defined here is the same as that defined in (3.39). The limiting case of the largest spatial scale state can be considered as the case $L_p \to \infty$, as $L_m \to \infty$, with $\varepsilon \equiv L_p/L_m \ll 1$. With this definition and the above consideration, we can always easily find a region near $m = n = 0$ in the WKB phase space, which can be as small as we wish. That is, for any given ε, there always exists a sufficiently small region near $m = n = 0$, for instance, $\gamma(|m| < \varepsilon^{\frac{1}{2}}, |n| < \varepsilon^{\frac{1}{2}})$. Outside this region, the WKB method is valid; inside this region, the WKB method is invalid. Therefore, although the point $m = n = 0$ in the WKB phase space is an invalid central wave number and the WKB method does not apply, there exists a sufficiently small region of width of $O(\varepsilon^{\frac{1}{2}})$, for example, outside this region, so that the analysis is valid.

The next step is to show that the dynamic behavior near the small region is the same as the behavior near $m = n = 0$. If this is the case, we can symbolically use the point $m = n = 0$ as a valid point in the analysis to describe the dynamic system.

Let

$$m = \varepsilon^{\frac{1}{2}} M, \qquad n = \varepsilon^{\frac{1}{2}} N; \tag{5.70}$$

then eqs. (5.26) and (5.27) become

$$\frac{\varepsilon^{\frac{1}{2}} D_g M}{DT} = \frac{\varepsilon^{\frac{1}{2}}}{F + \varepsilon M^2 + \varepsilon N^2} \{ k_0 M - \mu N - \varepsilon k_1 (M^2 + N^2) N \} \tag{5.71}$$

and

$$\frac{\varepsilon^{\frac{1}{2}} D_g N}{DT} = -\frac{\varepsilon^{\frac{1}{2}}}{F + \varepsilon M^2 + \varepsilon N^2} \{ \lambda M + k_0 N + \varepsilon k_0 (M^2 + N^2) M \}. \tag{5.72}$$

For the order of $O(\varepsilon^{\frac{1}{2}})$, we have

$$\frac{D_g M}{DT} = \frac{k_0 M - \mu N}{F} \tag{5.73}$$

and

$$\frac{D_g N}{DT} = -\frac{\lambda M + k_0 N}{F}. \tag{5.74}$$

The stability factor is readily found to be

$$p^2 = \frac{k_0^2 + \mu \lambda}{F^2}. \tag{5.75}$$

Compared with (5.30) to (5.32) in the text, we conclude that there exists a region of width of $O(\varepsilon^{\frac{1}{2}})$ and that the dynamic behavior near this region is the same as that near $m = n = 0$. This completes the proof.

For convenience, however, in the text, we will use $m = n = 0$ as a symbolically valid point with the understanding that the largest spatial scale state $m = n = 0$ is the limit case in which $L_p \to \infty$ as $L_m \to \infty$, with $L_p/L_m \ll 1$ so the WKB analysis is still valid.

REFERENCES

Andronov, A.A., and Pontryagin, A.A. (1937). Systemes Grossiers. *Dokl. Akad. Nauk. USSR* **14**, 247–251.

Andronov, A.A., and Witt, A. (1930). Sur la Thérie mathematiques des autooscillations. *C. Acad. Sci. Paris* **190**, 256–258.

Andronov, A.A., Vitt, E.A., and Khaiken, S.E. (1966). *Theory of Oscillators.* Pergamon Press, Oxford, Reprinted by Dover, New York, 1987.

Andronov, A.A., Leontovich, E.A., Gorden, I.I., and Maier, A.G. (1971). *Theory of Bifurcations of Dynamical Systems on a Plane.* Israel Program of Scientific Translations, Jerusalem.

Charney, J.B., and DeVore, J.G. (1979). Multiple flow equilibria in the atmosphere and blocking. *J. Atmos. Sci.* **36**, 1205–1216.

Curry, J.H. (1978). A generalized Lorenz system. *Communs. Math. Phys.* **60**, 193–204.

Elberry, R.L. (1968). A high-rotating general circulation model experiment with cyclic time changes. Atmospheric Science Paper No. 134. Colorado State University, Fort Collins.

Fultz, D., Long, R.R., Owens, G.V., Bohan, W., Kaylor, R., and Weil J. (1959). Studies of thermal convection in a rotating cylinder with some implications for large-scale atmospheric motions. *Meteorol. Monogr.*, No. 21.

Golubitsky, M., and Schaeffer, D.G. (1985). *Singularities and Groups in Bifurcation Theory.* Vol. I. Springer-Verlag, New York.

Golubitsky, M., Steward, I., and Schaeffer, D.G. (1988). *Singularities and Groups in Bifurcation Theory.* Vol. II. Springer-Verlag, New York.

Gruber, A. (1975). The wave number–frequency spectra of the 200-mb wind field in the tropics. *J. Atmos. Sci.* **32**, 1615–1625.

Guckenheimer, J., and Holmes, P. (1983). *Nonlinear Oscillations, Dynamical Systems, and Bifurcations of Vector Fields.* Springer-Verlag, New York.

Hart, J.E. (1984). A laboratory story of baroclinic chaos on the f-plane. *Tellus* **37A**, 286–296.

Hide, R. (1958). An experimental study of thermal convection in a rotating liquid. *Phil. Trans. Roy. Soc. London* **250A**, 441–478.

Hirsh, M.W., and Smale, S. (1974). *Differential Equations, Dynamical Systems and Linear Algebra.* Academic Press, New York.

Hopf, E. (1942). Abzweigung einer periodischen Lösung von einer stationären Lsung einer differential-systems. *Berl. Math.-Phys. Kl. Sächs. Acad. Wiss. Leipzig* **90**, 1–22. (An English translation of this paper and comments on it appear as Section 5 in Marsden and McCracken, 1976.)

Iooss, G., and Joseph, D.D. (1980). *Elementary Stability and Bifurcation Theory.* Springer-Verlag, New York.

Joseph, D.D. (1976). *Stability of Fluid Motions.* Vol. I, Springer-Verlag, New York.

Källen, E. (1982). Bifurcation properties of quasi-geostrophic, barotropic models and their relation to atmospheric blocking. *Tellus* **34**, 255–265.

Lorenz, E.N. (1963). Deterministic nonperiodic flow. *J. Atmos. Sci.* **20**, 130–141.

Lyapunov, M.A. (1893). *Comm. Soc. Mat. Kharkov.* (English translation in 1966: *Stability of Motion,* Academic Press, New York.)

Marsden, J.E., and McCracken, M. (1976). *The Hopf Bifurcation and Its Applications.* Springer-Verlag, New York.

McGuirk, J.P., and Reiter, E.R. (1976). A vacillation in atmospheric energy parameters. *J. Atmos. Sci.* **33**, 2079–2093.

Mitchell, K.E., and Dutton, J.A. (1981). Bifurcation from stationary to periodic solutions in a low-order model for forced, dissipative barotropic flow. *J. Atmos. Sci.* **38**, 690–716.

Moroz, I.M., and Holmes, P. (1984). Double Hopf bifurcation and quasi-periodic flow in a model for baroclinic instability. *J. Atmos. Sci.* **41**, 3147–3160.

Peixoto, M.M. (1962). Structural stability on two-dimensional manifolds. *Topology* **1**, 101–120.

Pfeffer, R.L., and Chiang, Y. (1967). Two kinds of vacillation in rotating laboratory experiments, *Mon. Wea. Rev.* **95**, 75–82.

Poincaré, H. (1892). *Les Méthodes Nouvelles de la Mécanique Céleste,* Vol. I. Gauthier-Villars, Paris.

Poincaré, H. (1899). *Les Méthodes Nouvelles de la Mécanique Céleste.* 3 vols. Gauthier-Villars, Paris.

Rouche, N., Habets, P., and Laloy, H. (1977). *Stability Theory by Liapunov Direct Method,* Springer-Verlag, New York.

Sparrow, C. (1982). *The Lorenz Equations: Bifurcation, Chaos, and Strange Attractors.* Springer-Verlag, New York.

Tung, K.K., and Rosenthal, A.J. (1985). Theories of multiple equilibria—A critical reexamination, Part I: Barotropic models, *J. Atmos. Sci.* **42**, 2804–2819.

Vickroy, J.G., and Dutton, J.A. (1979). Bifurcation and catastrophe in a simple, forced, dissipative quasi-geostrophic flow. *J. Atmos. Sci.* **36**, 42–52.

Webster, P.J., and Keller, J.L. (1975). Atmospheric variations: Vacillations and index cycles. *J. Atmos. Sci.* **32**, 1283–1300.

Weng, H., Barcilon, A., and Magnan, J. (1986). Transitions between baroclinic flow regimes. *J. Atmos. Sci.* **43**, 1760–1777.

Winn-Nielsen, A. (1979). Steady states and stability properties of a low-order barotropic system with forcing and dissipation. *Tellus* **31**, 375–386.

Yang, H. (1988a). Global behavior of the evolution of a Rossby wave packet in barotropic flows on the earth's δ-surface. *J. Atmos. Sci.* **45**, 133–146.

Yang, H. (1988b). Bifurcation properties of the evolution of a Rossby wave packet in barotropic flows on the earth's δ-surface. *J. Atmos. Sci.* **45**, 3667–3683.

Yoden, S. (1985). Bifurcation properties of a quasi-geostrophic barotropic, low-order model with topography. *J. Meteorol. Soc. Japan* **63**, 535–546.

6

Secondary Bifurcation

6.1 Introduction

In Chapter 5 we analytically investigated the bifurcation properties of the evolution of the wave packet in the symmetric basic flow due to symmetric and asymmetric topographies, using the Rossby wave packet approximation, δ-surface approximation of the earth's surface, and the WKB method. The results show that the topological structure of the evolution of a Rossby wave packet varies with basic flows and the topography. The subcritical and supercritical bifurcations, as well as the reverse subcritical and supercritical bifurcations, were found analytically. The effect of a zonal basic flow on the bifurcation differs from that of a meridional basic flow. The mixed scale equilibrium states were only found associated with the asymmetric topography.

However, in earlier studies, we discussed the effects of symmetric basic flow on the bifurcation properties and only found the primary bifurcation. In real geophysical fluids, however, the asymmetric basic flow is often important; for example, the southwesterly jet and the southeasterly jet are often observed in Asia. Moreover, both the calculated results and the laboratory experiments strongly suggest that there is a sequence of bifurcations in geophysical flows as the parameter increases (Lorenz, 1963; Pedlosky and Frenzen, 1980; Weng et al., 1986; Hart, 1984 in the laboratory experiment). These studies also demonstrated a reverse bifurcation cascade. Therefore, a further investigation of the bifurcation properties of the evolution of a wave packet, especially where the asymmetric basic flow is included, is worthwhile, in order to better understand the bifurcation properties and the effects of asymmetric basic flow and topography in geophysical flows. This is the topic of the present chapter (Yang, 1988c).

6.2 Secondary and Cascading Bifurcation Theory

The sequence of bifurcations has been investigated in many studies. For example, thermal convection experiments (Krishnamurti, 1970, 1973; Gollub and Benson, 1980) suggest that the transition from laminar conduction to turbulent convection occurs by a sequence of distinct transitions. At each transition, the fluid assumes a more complex spatial and/or temporal pat-

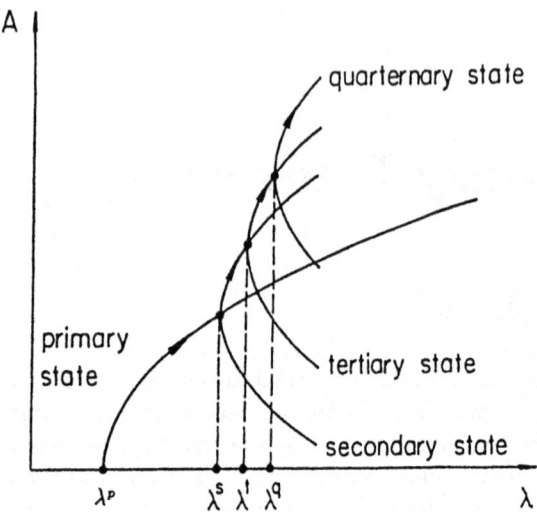

FIGURE 6.1. The response diagram illustrating a cascading bifurcation, where λ is the bifurcation parameter and A is the equilibrium state of the solutions. The values λ^p, λ^s, λ^t, and λ^q denote primary, secondary, tertiary, and bifurcation points, respectively.

tern, in accordance with Landau's conjecture (Landau, 1944; Landau and Lifschitz, 1959, 1987). In addition, this sequence of bifurcations has been observed in buckling elastic plates and in chemical and biochemical instabilities (e.g., Nicolis and Prigogine, 1977) related to morphogenesis and pattern formation. These transitions occur by a sequence of bifurcations as a bifurcation parameter, λ, is increased.

In the bifurcation theory, in general, if the second equilibrium state is bifurcated from the first equilibrium state, which usually exists for all ranges of the bifurcation parameter, as the bifurcation parameter reaches the first critical value of parameter, a bifurcation called the *primary bifurcation,* such as is described in Chapter 5, is found. The first critical value of the parameter is called the *primary bifurcation value* or the *primary bifurcation point.* If a third equilibrium state is bifurcated from the second equilibrium state, as the parameter reaches the second critical value, then this bifurcation is called the *secondary bifurcation.* The second critical value of the parameter is called the *secondary bifurcation value,* accordingly. Similarly, we might have tertiary and quarternary bifurcations and tertiary and quarternary bifurcation values.

Therefore, there might be a sequence of bifurcations. In the terminology of bifurcation theory, this has been called the *cascading bifurcation* (Reiss, 1983; Erneux and Reiss, 1983). Figure 6.1 illustrates this cascading bifurcation, where λ^p, λ^s, λ^t, and λ^q correspond to primary, secondary, tertiary, and quarternary bifurcation points, respectively. That is to say, there occur primary, secondary, tertiary, and quarternary bifurcations, respectively at λ^p, λ^s, λ^t, and λ^q.

In the following, we show that, in the evolution of a wave packet on one side of an asymmetric basic flow, the evolution of the wave packet will exhibit a cascading bifurcation with an increasing bifurcation parameter (the topography parameter) and a reverse cascading bifurcation with a decreasing bifurcation parameter, as the bifurcation parameter, that is, the topography parameter is varied. When the topography parameter reaches the first critical value $\lambda = \lambda^p = 0$, the primary bifurcation occurs with four bifurcated, purely scale states. As the topography is further increased to the second critical value $\lambda = \lambda^s = \lambda_c$, there appears a secondary bifurcation with mixed scale states. It is possible for transitions to occur among the three kinds of wave packet structural vacillations. However, in this study, we did not find the tertiary and quaternary bifurcations, as shown in Figure 6.1.

6.3 Primary Bifurcation

The governing equations in this chapter are the same as those in earlier chapters. In the case when $k_0 = 0$ (the symmetric topography case), eqs. (3.49) and (3.50) can be rewritten as follows:

$$\frac{D_g m}{DT} = -\frac{n}{K^2}\{\mu + k_1(m^2 + n^2)\} \tag{6.1}$$

and

$$\frac{D_g n}{DT} = -\frac{m}{K^2}\{\lambda + k_2(m^2 + n^2)\}, \tag{6.2}$$

where only the main part of the effect of basic flow has been considered, and k_1, k_2, λ, and μ are defined by

$$\lambda = \delta_0 - \frac{\partial^2 \eta_B}{\partial y \partial Y}, \tag{6.3a}$$

$$\mu = \frac{\partial^2 \eta_B}{\partial x \partial X}, \tag{6.3b}$$

$$k_1 = \frac{\partial V}{\partial X}, \tag{6.3c}$$

and

$$k_2 = \frac{\partial U}{\partial Y}. \tag{6.3d}$$

The physical values of these parameters are found in Chapter 5, and their importance to the study of geophysical fluids is discussed.

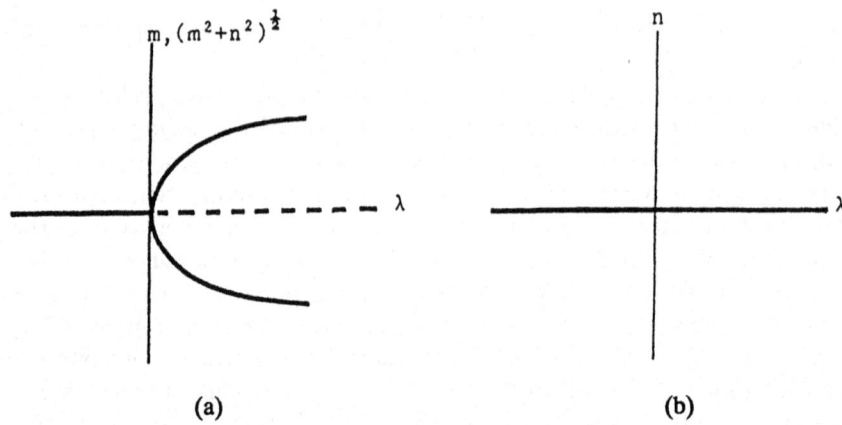

FIGURE 6.2. The bifurcation diagram in an asymmetric basic flow on concave topography. The wave packet is located on the left-hand side of a southwesterly jet. (a) The longitudinal scale states and the whole spatial scale states and (b) the latitudinal scale states. Key: - - -, unstable equilibrium states; —, neutral or stable equilibrium states.

In what follows, we take λ as the main bifurcation parameter where λ describes the joint effect of the δ-effect and the curvature of topography along the north–south direction. In the following, concave topography means $\mu > 0$ and convex topography means $\mu < 0$, for convenience. In this section, we consider cases such that $\mu k_1 > 0$. Examples of such cases could be where the Rossby wave packet is located on (1) the left-hand side of a southwesterly jet and concave topography; (2) the right-hand side of a southwesterly jet and convex topography; (3) the right-hand side of a southeasterly jet and concave topography; and (4) the right-hand side of a southeasterly jet and convex topography. In these cases, there are two kinds of equilibrium states in the system, that is,

$$m = 0, \qquad n = 0 \tag{6.4}$$

and

$$m^2 = -\frac{\lambda}{k_2}, \qquad n = 0. \tag{6.5}$$

The equilibrium state (6.4) corresponds to the largest spatial scale state, whereas (6.5) corresponds to the two purely longitudinal spatial scale states. It should be pointed out that although the point $m = n = 0$ is not a valid center wave number for the wave packet, we can always replace the point $m = n = 0$ with a region of width of $O(\varepsilon^{1/2})$, where ε is defined by $\varepsilon = O(L/a) = L_p/L - m \ll 1$, which is the same as defined in Chapter 5. Here, L_p is the length scale of wave packet, whereas L_m is the length scale of variations in the medium. Outside this region of the WKB phase space, the WKB analysis is valid. The equilibrium state $m = n = 0$ should

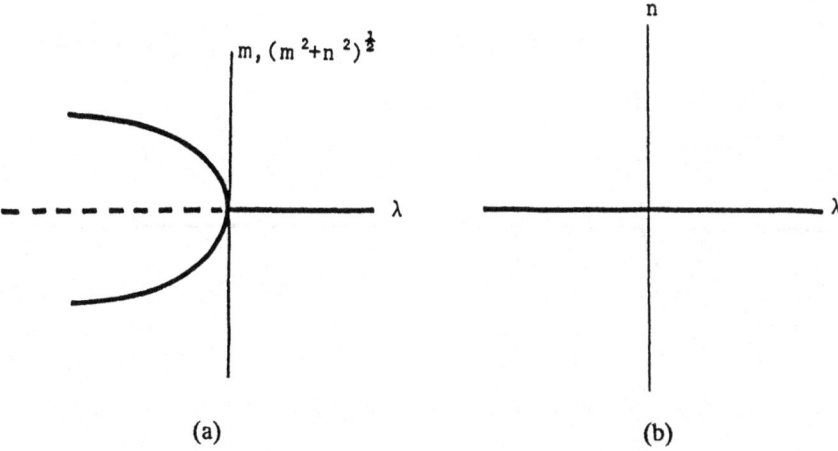

FIGURE 6.3. The bifurcation diagram for the case in which the wave packet is located on the right-hand side of a southwesterly jet on convex topography. (a) The longitudinal scale states and the whole spatial scale states and (b) the latitudinal scale states. Key: - - -, unstable equilibrium states; —, neutral or stable states.

be understood as the limiting case in which $L_p \to \infty$ as $L_m \to \infty$, with $\varepsilon = L_p/L_m \ll 1$. For convenience, we still symbolically treat the point $m = n = 0$ in the WKB phase space as a valid point. A detailed justification of this largest spatial scale state is given in Appendix A in Chapter 5 or Yang (1988b).

These results are similar to the case of symmetric basic flow discussed in Chapter 5. The stability analysis shows that the largest spatial scale state is unstable when $\mu\lambda > 0$, whereas the purely longitudinal scale states are unstable when $\lambda\left(\frac{k_1}{k_2}\lambda - \mu\right) > 0$. For instance, equilibrium states (6.5) have the spatial scale with wave number 5, when the wave packet is located on the right-hand side of the southwesterly jet and convex topography, as was used before. This spatial scale is a typical one for a synoptic disturbance system in the atmosphere.

When the wave packet is located on the left-hand side of a southwesterly jet with a concave topography, the largest spatial scale state changes its stability as the topography parameter crosses the critical value, that is, zero. The purely longitudinal scale states appear with neutral stabilities. In Figure 6.2 and the following bifurcation diagrams, we again use the local wave numbers in the X- and Y-direction to represent the longitudinal spatial scale and the latitudinal spatial scale, respectively. The total local wave number is defined by $(m^2 + n^2)^{\frac{1}{2}}$ to represent the whole spatial scale of the Rossby wave packet because of the special relationship between the wave number and the spatial scale. Figure 6.2a illustrates the bifurcation diagram for the longitudinal scale states and the whole spatial scale states,

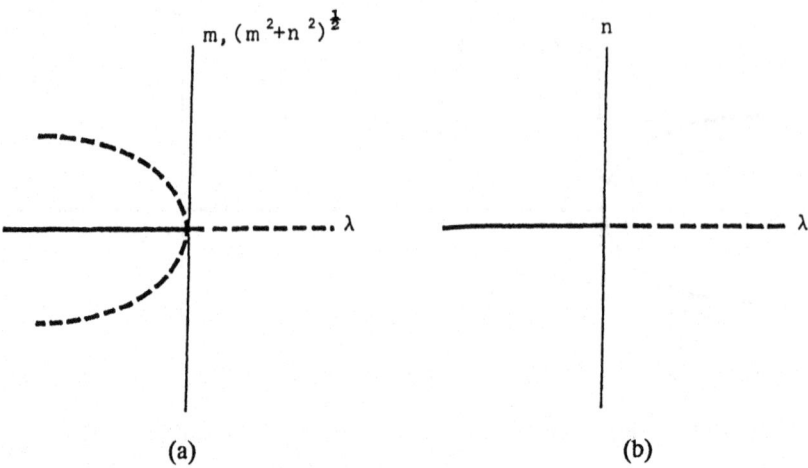

FIGURE 6.4. The bifurcation diagram for the case in which the wave packet is located on the left-hand side of a southeasterly jet on a concave topography. (a) The longitudinal scale states and the whole spatial scale states and (b) the latitudinal scale states. Key: - - -, unstable equilibrium states; —, neutral or stable equilibrium states.

whereas Figure 6.2b corresponds to the latitudinal spatial scale state, with unstable and neutral or stable equilibrium states. From Figure 6.2, we find that, in this case, the bifurcation is supercritical.

Figure 6.3 illustrates a wave packet located on the right-hand side of a southwesterly jet on convex topography. From the figure, one finds that the largest spatial scale state is unstable when the topography parameter is less than its critical value. Since the wave packet is located on the right-hand side of a southwesterly jet, the purely longitudinal spatial scale states occur only on the left half of the topography parameter. The purely longitudinal scale states have neutral stabilities. Nevertheless, the results are the opposite of that illustrated in the previous figure (Figure 6.2a). Therefore, the bifurcation is the reverse supercritical bifurcation.

The case in which the Rossby wave packet is located on the left-hand side of a southeasterly jet with a concave topography is shown in Figure 6.4. From this figure, one readily finds that the largest spatial scale state is unstable when the topography parameter is greater than its critical value, whereas the purely longitudinal scale states occur when $\lambda > 0$. The bifurcation in this case is thus the subcritical bifurcation. Figure 6.5 corresponds to the case in which the Rossby wave packet is located on the right-hand side of a southeasterly jet on convex topography. From the figure, it is easily seen that the bifurcation is the reverse subcritical bifurcation. Figures 6.3a, 6.4a, and 6.5a correspond to the bifurcation diagrams for the longitudinal scale and the whole spatial scale states, whereas Figures 6.3b, 6.4b, and

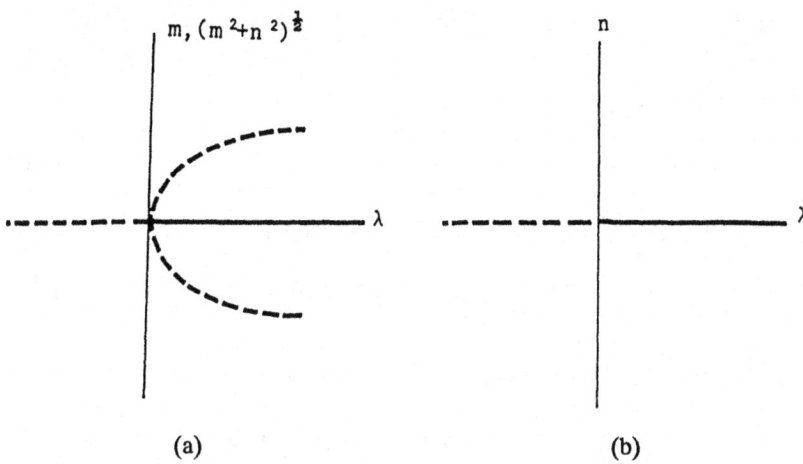

FIGURE 6.5. The bifurcation diagram for the case in which the wave packet is located on the right-hand side of a southeasterly jet on convex topography. (a) The longitudinal and the whole spatial scale states and (b) the latitudinal scale states. Key: - - -, unstable equilibrium states; —, neutral or stable states.

6.5b are for the latitudinal scale states.

Although the asymmetry of the basic flow has been included in these cases, comparison with the results in Chapter 5, or Yang (1988b), shows that the bifurcation properties in these cases are similar to those in the case in which only the symmetric basic flow has been considered. Nevertheless, the bifurcation properties will differ in the other cases, since there the secondary bifurcation appears. Therefore, the cascading bifurcation is found. This will be the topic of the following section.

6.4   Secondary Bifurcation

When we consider cases in which $\mu k_1 < 0$, we gain further insight into the bifurcation properties of the evolution of a wave packet. The cascading bifurcation occurs when the topography parameter increases so that there is a sequence of bifurcations. When the parameter reaches the first critical value $\lambda = \lambda^p = 0$, the primary bifurcation occurs. As the parameter is further increased to the second critical value $\lambda = \lambda^s = \lambda_c$, the secondary bifurcation will appear, which implies the existence of the cascading bifurcation.

This condition $\mu k_1 < 0$ is satisfies in cases in which the Rossby wave packet is located on, for example, (1) the right-hand side of a southeasterly jet on concave topography; (2) the right-hand side of a southwesterly jet on concave topography; (3) the left-hand side of a southwesterly jet on convex topography; and (4) the left-hand side of a southeasterly jet on convex topography. In such cases, there are three kinds of possible equilibrium states, which can be readily derived from (6.1) and (6.2), as follows:

A. The largest spatial scale state, that is,

$$m = 0, \qquad n = 0; \tag{6.6}$$

B. (1): the purely latitudinal scale states, that is,

$$m = 0, \qquad n^2 = N^2 = -\frac{\mu}{k_1}; \tag{6.7}$$

(2): the purely longitudinal scale states, that is,

$$m^2 = M^2 = -\frac{\lambda}{k_2}, \qquad n = 0; $$

C. the mixed scale states, that is,

$$m^2 = -\frac{\lambda}{k_2} - n^2, \qquad n \text{ arbitrary,} \tag{6.8}$$

if and only if

$$\lambda = \lambda^s = \lambda_c \equiv \frac{\mu}{k_1} k_2. \tag{6.9}$$

It should be noted that the mixed scale state only exists at the critical value for the parameter λ. Thus, λ^s corresponds to the secondary bifurcation point. The case, for example, on the left-hand side of a southwesterly jet with shears of about $2\,\text{m/s}$ in the U-component and $1\,\text{m/s}$ in the V component, $100\,\text{km}$ out from the jet center, and in which a convex topography has the distribution given by $\eta_B = H - 0.5(xX + yY)$, corresponds to that case in which the purely latitudinal scale states have the spatial scale with wave number 7. The purely longitudinal scale states have the spatial scale with wave number 5. The mixed spatial scale states also have the spatial scale with wave number 5. These spatial scales are indeed the typical scales for the synoptic disturbance systems in the atmosphere. The secondary critical bifurcation value is about order $O(1)$, which is a physically reasonable value.

The linear stability analysis shows that about the largest spatial scale state (6.6), the stability factor p_L is determined by

$$p_L^2 = \frac{\mu\lambda}{F^2}. \tag{6.10}$$

About the purely latitudinal scale states (6.7) the stability factor p_N is determined by

$$p_N^2 = -\frac{2\mu}{K_N^2}\left(\lambda - \frac{k_2}{k_1}\mu\right), \tag{6.11}$$

where

$$K_N^2 = F - \frac{\mu}{k_1}. \tag{6.12}$$

About the purely longitudinal scale states (6.7b), the stability factor p_M is determined by

$$p_M^2 = -\frac{2\lambda}{K_M^4}\left(\mu - \frac{k_1}{k_2}\lambda\right), \tag{6.13}$$

where

$$K_M^2 = F - \frac{\lambda}{k_2}. \tag{6.14}$$

About the mixed scale states (6.8), we have the following linearized equations to determine the stability:

$$\frac{D_g m}{DT} = -\frac{2Nk_1}{K_{MN}^2}(Mm + Nn) \tag{6.15a}$$

and

$$\frac{D_g n}{DT} = -\frac{2Mk_2}{K_{MN}^2}(Mm + Nn), \tag{6.15b}$$

where

$$K_{MN}^2 = F - \frac{\mu}{k_1}. \tag{6.16}$$

Here m and n in (6.15a,b) are perturbations about the equilibrium states; the primes m and n have been dropped; and M and N are the equilibrium values for the m and n, respectively, determined by (6.8).

The characteristic equation for (6.15a) and (6.15b) could be easily derived as

$$p_{MN}^2 + \frac{2MN}{K_{MN}^2}(k_1 + k_2)p_{MN} = 0. \tag{6.17}$$

From (6.17) one could conclude that the stabilities of the mixed scale states are determined by $MN(k_1+k_2)$, where p_{MN} is not equal to zero. Therefore, in Quadrants I and III, the mixed spatial scale states are neutrally stable when $(k_1 + k_2) > 0$, whereas the mixed spatial scale states are unstable when $(k_1+k_2) < 0$. In quadrants II and IV, however, the mixed spatial scale states are neutrally stable when $(k_1 + k_2) < 0$, whereas they are unstable when $(k_1 + k_2) > 0$.

The bifurcation diagrams for each case are shown in Figures 6.6 through 6.9. Points A and B are the mixed scale states. The results show that there is a cascading bifurcation, as was also found in other studies (e.g., Reiss, 1983; Erneux and Reiss, 1984). However, this cascading bifurcation is of a special type, since the secondary bifurcation is a transcritical bifurcation. Figure 6.6 illustrates the cascading bifurcation in such a case in which the wave packet is located in a region where $k_1 = (\partial V/\partial X) < 0$ and $k_2 = (\partial U/\partial Y) < 0$ on concave topography. Figure 6.6a is the bifurcation diagram for the

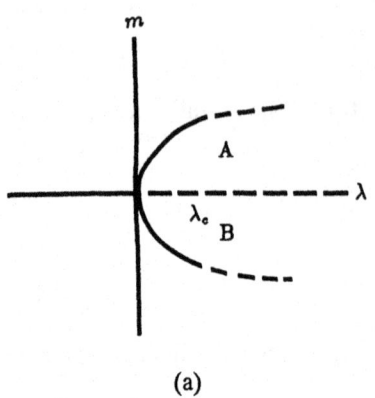

(a)

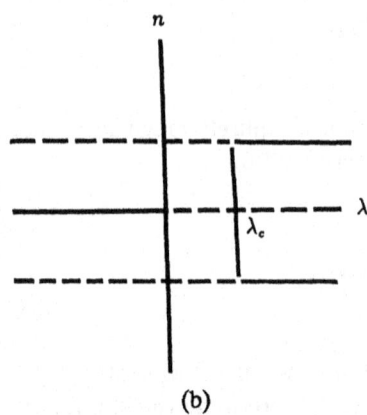

(b)

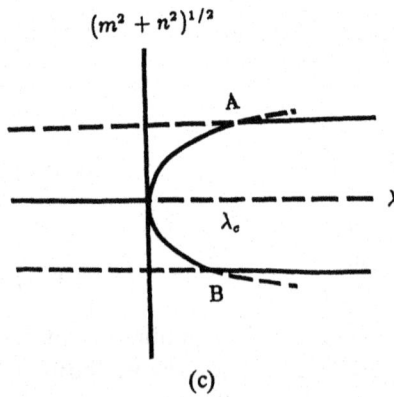

(c)

FIGURE 6.6. The bifurcation diagram for the case in which the wave packet is located in a region where the basic flow satisfies $k_1 < 0$ and $k_2 < 0$ on concave topography, where λ_c denotes the secondary bifurcation point. (a) The longitudinal scale states, (b) the latitudinal scale states, and (c) the whole spatial scale states. Key: - - -, unstable equilibrium states; —, neutral or stable equilibrium states.

longitudinal scale states. There are three possible equilibrium states: the largest spatial scale state, the purely longitudinal scale state, and the mixed scale states at $\lambda = \lambda^s = \lambda_c$ denoted by A and B for a certain latitudinal scale's wave packet. The stabilities of the purely longitudinal scale states change at the secondary bifurcation point. Figure 6.6b corresponds to the latitudinal scale bifurcation diagram, in which there are also three possible equilibrium states: the largest spatial scale state, the purely latitudinal scale states, and the mixed scale states at $\lambda = \lambda^s = \lambda_c$. Again, the stabilities of the purely latitudinal scale states change at the secondary bifurcation point. Here the straight line at $\lambda = \lambda^s = \lambda_c$ corresponds to the mixed scale state, which only exists in the certain range of latitudinal scale, that is, $n^2 \leq -\lambda/k_2$. When n^2 reaches N^2, it will become the purely latitudinal scale state.

The bifurcation diagram for the whole spatial scale of a wave packet is shown in Figure 6.6c. From this diagram, one finds that as the bifurcation parameter λ increases, the wave packet at first has the largest spatial scale as its equilibrium state. As the bifurcation parameter reaches the first critical bifurcation value, that is, the primary bifurcation point, a pri-

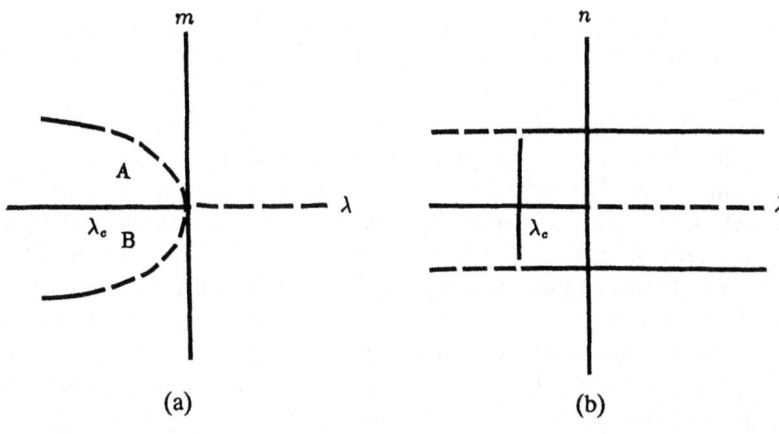

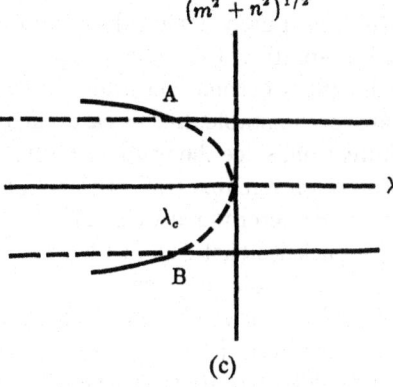

FIGURE 6.7. The bifurcation diagram for the case in which the wave packet is located on the right-hand side of a southwesterly jet on concave topography. (a) The longitudinal scale states, (b) the latitudinal scale states, and (c) the whole spatial scale states. Key: - - -, unstable equilibrium states; —, neutral or stable equilibrium states.

mary bifurcation occurs and the solution bifurcates into purely longitudinal scale states. The primary bifurcation is the supercritical bifurcation. Furthermore, as the bifurcation parameter increases to the next critical value λ_c, defined by (6.9), which is the secondary bifurcation point, the wave packet will exhibit a secondary bifurcation. These states correspond to the mixed scale states denoted by A and B in the figure. The stabilities of the mixed scale states in this case are determined by (6.17). Finally, as the topography parameter further increases, the wave packet has two purely latitudinal states. The secondary bifurcation is a transcritical bifurcation. From Figure 6.6c, one finds that there is a sequence of bifurcations as the bifurcation parameter increases. Therefore, the cascading bifurcation has been found. In the present model, however, we did not find tertiary or quarternary bifurcation, which can be found in some other physical models (e.g., Reiss, 1983; Erneux and Reiss, 1983).

In the case in which the wave packet is located on the right-hand side of a southwesterly jet with a concave topography, the bifurcation properties are different. The results are summarized in Figure 6.7. The bifurcation diagrams for the longitudinal scale and latitudinal scale are shown in Figures

6.7a and b, respectively. Figure 6.7c is the bifurcation diagram for the whole spatial scale state. In this case, the mixed scale states are stable, denoted by A and B in Figures 6.7a and c, and the solid line parallels to the n-axis in Figure 6.7b. These figures show that when the topography parameter is greater than zero, namely, the primary bifurcation value, there are three equilibrium states. The largest spatial scale state is unstable, whereas the two purely latitudinal scale states are stable. As the bifurcation decreases to the primary bifurcation point, the primary bifurcation occurs and the largest spatial scale changes its stability. Two additional purely longitudinal scale states occur when the parameter crosses the primary bifurcation point. These additional states are unstable.

When the parameter further decreases to the second critical value, the secondary bifurcation occurs and the mixed spatial scale states appear, denoted by points A and B in Figures 6.7a and c and the solid line in Figure 6.7b. The purely longitudinal and purely latitudinal scale states change their stabilities at the secondary bifurcation point. The purely longitudinal scale states become neutrally stable, whereas the purely latitudinal scale states are unstable. In this case, the primary bifurcation is now a subcritical bifurcation, since the primary bifurcated states are unstable. The secondary bifurcation remains transcritical. The stabilities of the mixed spatial scale states are determined by (6.17). It will be verified in the next section that wave packet structural vacillations will occur on almost the entire domain of the WKB phase space.

The case in which the wave packet is located on the left-hand side of a southwesterly jet on convex topography is illustrated in Figure 6.8. The results reveal that when the wave packet is on convex topography, the bifurcation is a reverse supercritical bifurcation if the packet is located on the left-hand side of a southwesterly jet; whereas the bifurcation is a reverse subcritical one if the wave packet is located in a region with $k_1 > 0$ and $k_2 > 0$ on convex topography. The cascading bifurcation has also been found in Figures 6.7 to 6.9. At the secondary bifurcation point λ^s, the mixed spatial scale state is neutrally stable in Figure 6.8 and unstable in Figure 6.9. The reverse cascading bifurcation occurs for the case shown in Figure 6.9. At the secondary bifurcation point, the stable mixed scale state is a time periodic state, which corresponds to the wave packet vacillation. Moreover, the evolution of a wave packet at the secondary bifurcation is always time periodic on the entire domain of the WKB phase space when the mixed spatial scale states are neutrally stable.

In the previous section, we found only the primary bifurcation. However, if we consider the basic flow to be a varying parameter, the results suggest that, for the same type of topography, the wave packet bifurcation on one side of the asymmetric basic flow will differ from that on the other side. The secondary bifurcation, and hence the cascading bifurcation, only exists on one side of an asymmetric basic flow. Therefore, on the same type of topography, the topological structure of the evolution of a wave packet

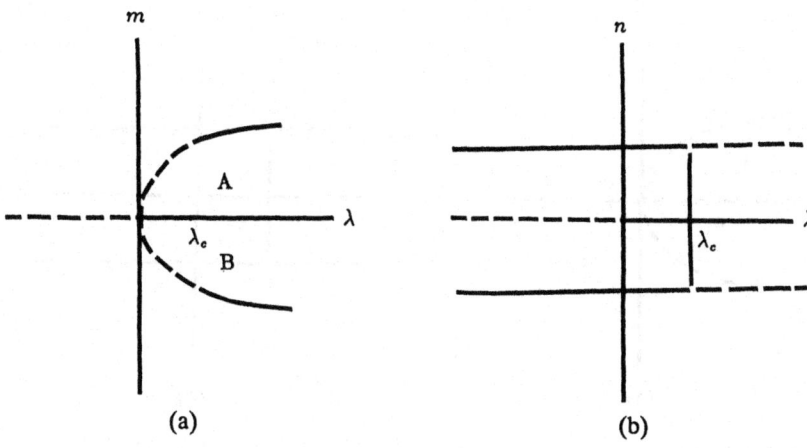

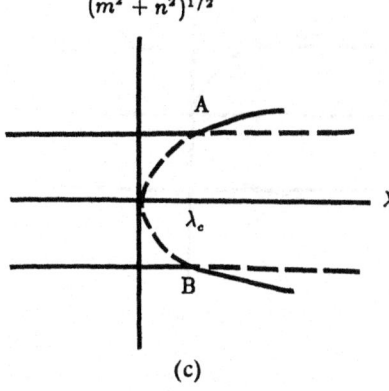

FIGURE 6.8. The bifurcation diagram for the case in which the wave packet is located on the left-hand side of a southwesterly jet on convex topography. (a) The longitudinal scale states, (b) the latitudinal scale states and (c) the whole spatial scale states. Key: - - -, unstable equilibrium states; —, neutral or stable states.

in the WKB phase space differs from one side of the basic flow to the other side. From the topological point of view, on the one hand, the system governing the evolution of a wave packet is structurally unstable in the center of an asymmetric basic flow. From a physical point of view, on the other hand, a wave packet will evolve differently from one side of an asymmetric basic flow to another. Therefore, when a Rossby wave packet passes the center of an asymmetric basic flow, the bifurcation will also occur, which is expected in real geophysical flows. The results show that both the topography and the basic flow play very important roles in the bifurcation properties of geophysical fluids. In order to better understand the bifurcation properties discussed above, in the following section we use the WKB phase space to see what it will look like in such a phase space, as was done in earlier chapters.

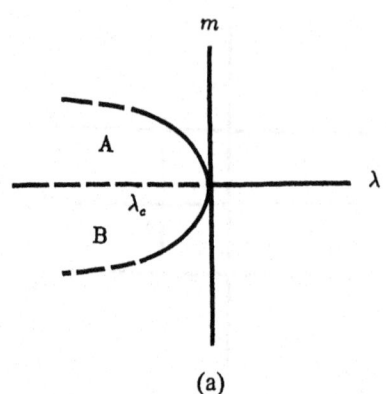

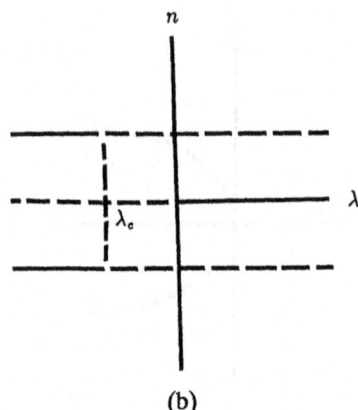

(a)   (b)

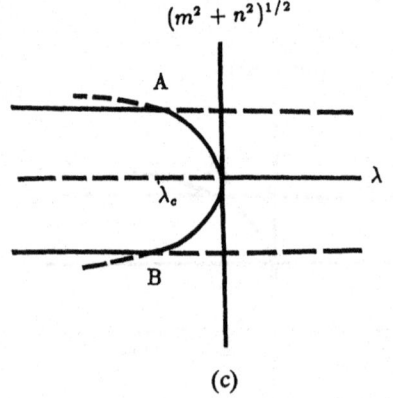

(c)

FIGURE 6.9. The bifurcation diagram for the case in which the wave packet is located in a region where the basic flow satisfies $k_1 > 0$ and $k_2 > 0$ on convex topography. (a) The longitudinal scale states, (b) the latitudinal scale states, and (c) the whole spatial states. Key: - - -, unstable equilibrium states; —, neutral or stable states.

6.5 Various Trajectories in the WKB Phase Space: Three Kinds of Wave Packet Structural Vacillation

Our use of the WKB phase space helps us to see the evolution of such a wave packet under various conditions clearly. These are discussed from the geometric point of view. Therefore, in what follows, we consider the WKB trajectories for different topography parameters in the WKB phase space. Special attention is given to the bifurcation properties associated with the structural vacillation in the wave packet.

It could be easily found from (6.1) and (6.2) that the divergence of the WKB phase space is

$$\frac{\partial}{\partial m}\left(\frac{D_g m}{DT}\right) + \frac{\partial}{\partial n}\left(\frac{D_g n}{DT}\right) = \frac{2mn}{K^4}\{(\lambda + \mu) - F(k_1 + k_2)\}. \qquad (6.18)$$

Equation (6.18) tells us that the divergence does change its sign in the phase space. Thus, closed orbits may exist in the WKB phase space, which indicates the occurrence of the wave packet's structural vacillation. Wave packet structural vacillation has been discussed and shown, in earlier chapters and also by Yang (1988a), to be similar to the wave vacillation found in geophysical fluids (e.g., Hide, 1958; Fultz et al., 1959; Pfeffer and Chiang, 1967; Elberry, 1968; McGuirk and Reiter, 1976; Gruber, 1975; Webster and Keller, 1975). However, from the theory of planer dynamic system (e.g., Andronov et al., 1966), we know that for a system such as the present one, all possible nonwandering sets will always fall into the following three classes: (1) fixed points, (2) closed orbits, and (3) the unions of fixed points and the trajectories connecting the fixed points. The latter are called *heteroclinic orbits* when they connect distinct points and *homoclinic orbits* when they connect a point to itself (e.g., Guckenheimer and Holmes, 1983; Sparrow, 1982).

As examples, we consider two cases. The first case is the one in which the wave packet on concave topography is located on the right-hand side of a southwesterly jet. The bifurcation diagrams for this case are shown in Figure 6.7. We consider the special case, with $k_2 = -k_1$. A direct substitution shows that, in this case, the system has the following common integral:

$$k_1(m^2 + n^2)^2 + 2(\mu n^2 - \lambda m^2) = C, \qquad (6.19)$$

where C is a constant of integration. When the parameter λ is greater than the primary bifurcation value, that is, zero, we consider a case such as

$$\lambda = \mu = -a^2 k_1, \qquad (6.20)$$

where a is a positive constant. Thus, eq. (6.19) becomes

$$(m^2 + n^2)^2 - 2a^2(n^2 - m^2) = C', \qquad (6.21)$$

where C' is a constant of integration. If we write $C' = C^4 - a^4$, where C is a positive constant, we obtain

$$(m^2 + n^2)^2 - 2a^2(n^2 - m^2) = C^4 - a^4. \qquad (6.22)$$

The curves (6.22) constitute a family of Cassini ovals, in which the foci are $A(0, a)$ and $B(0, -a)$. An elementary analysis shows that for $C > a\sqrt{2}$, the curves (6.22) are *convex* ovals; for $a < C < a\sqrt{2}$, they are *pinched* ovals; and for $C = a$, the curve is a *lemniscate*. For $0 < C < a$, the curves break into two separate ovals and finally for $C = 0$, they degenerate into two points, A and B. Each of the ovals is a path or trajectory of the evolution of a wave packet. The lemniscate consists of three paths: the saddle point $O(0, 0)$ and two separatrices, which form loops. The lemniscate obviously is the homoclinic orbit of the system. The points A and B are centers. The flow direction of the trajectories in the WKB phase space is readily defined by the sign of $D_g n/DT$. We thus obtain the configuration shown in Figure 6.10 in the WKB phase space.

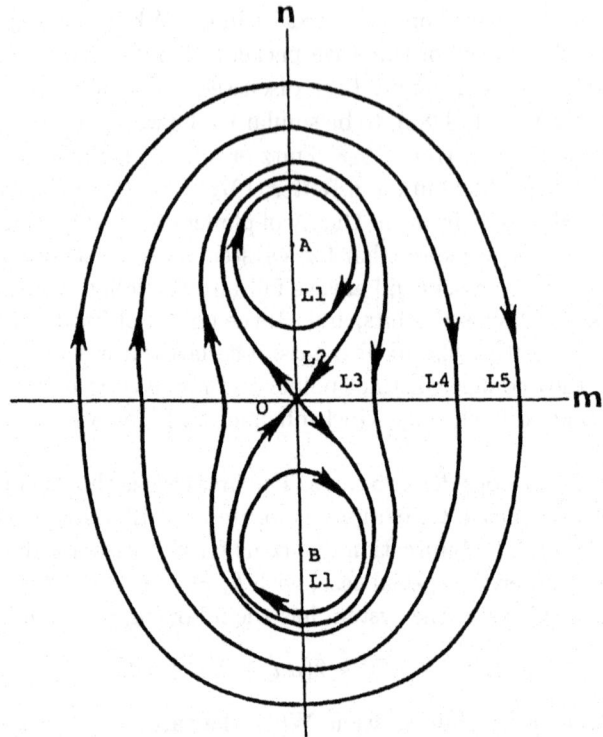

FIGURE 6.10. The WKB phase space for the case in which the wave packet is located on the right-hand side of a southwesterly jet on concave topography when the bifurcation parameter is greater than the primary bifurcation value. $A(0, N)$ and $B(0, -N)$ are centers; $O(0,0)$ is a saddle point.

In Figure 6.10, L1, L2, L3, L4, and L5 correspond, respectively, to

L1: $C < a$;

L2: $C = a$;

L3: $a < C < a\sqrt{2}$;

L4: $C = a\sqrt{2}$; and

L5: $C > a\sqrt{2}$.

As discussed, L3, L4, and L5 correspond to structural vacillations in the wave packet. Nevertheless, L1 describes a special kind of wave packet structural vacillation. This kind of packet structural vacillation is more easily observed in the annulus and real geophysical flows, since the local wave number in the Y-direction never becomes zero, which means that the tilt of the system will never parallel the east–west direction. This kind of wave packet vacillation is restricted to a range of longitudes.

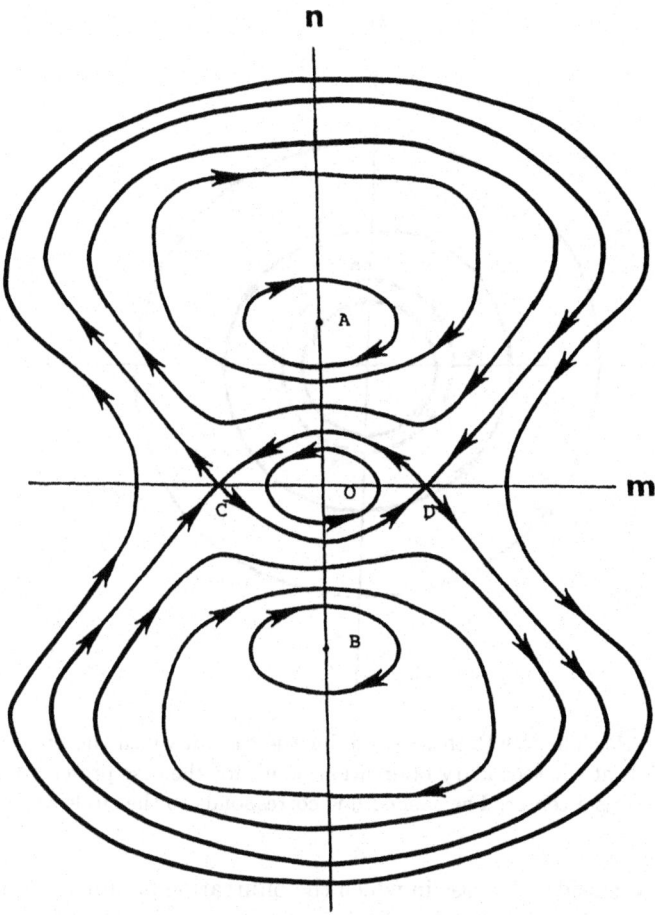

FIGURE 6.11. The WKB phase space for the case in which the topography parameter λ is less than the primary bifurcation value, that is, zero, and greater than the secondary bifurcation value λ_c, for the case presented in Figure 6.10. $A(0, N)$, $O(0, 0)$, and $B(0, -N)$ are centers; $C(-M, 0)$ and $D(M, 0)$ are saddle points.

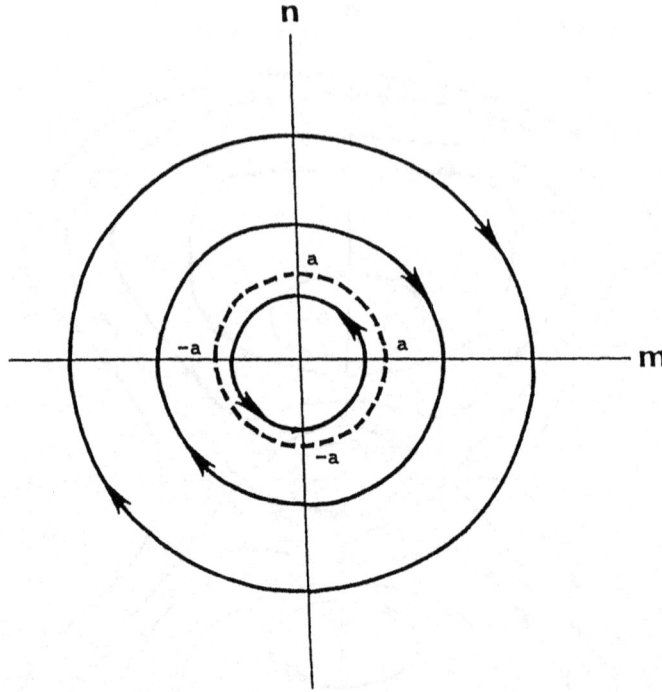

FIGURE 6.12. The WKB phase space for the case in which the bifurcation parameter λ is at the secondary bifurcation point, for the case presented in Figure 6.10. $O(0,0)$ is a center. The dashed line corresponds to the circle $m^2 + n^2 = a^2$.

We now consider the case in which the bifurcation parameter λ decreases beyond the primary bifurcation point, but not yet at the secondary bifurcation point. In this case, we still have the common integral (6.19). Purely longitudinal spatial scale states are seen. These purely longitudinal spatial scale states, however, are unstable, whereas the largest spatial scale state and the purely latitudinal spatial scale states are neutrally stable. From Figure 6.11 we find that there are three centers at $O(0,0)$, $A(0,N)$, and $B(0,-N)$, and two saddle points at $C(-M,0)$ and $D(M,0)$ in the WKB phase space. The centers $O(0,0)$, $A(0,N)$, and $B(0,-N)$ correspond to the largest spatial scale state and the two purely latitudinal spatial scale states, whereas the two saddle points $C(-M,0)$ and $D(M,0)$ correspond to the two purely longitudinal spatial scale states. The unions connecting the saddle points are heteroclinic orbits, which constitute the so-called *homoclinic cycle* (Guckenheimer and Holmes, 1983). The homoclinic cycle separates the wave packet structural vacillations. When the topography parameter λ reaches the secondary bifurcation point λ_c, the common integral (6.19) still holds. If we let $\mu = -\lambda = -a^2 k_1$, where a is a positive constant, then

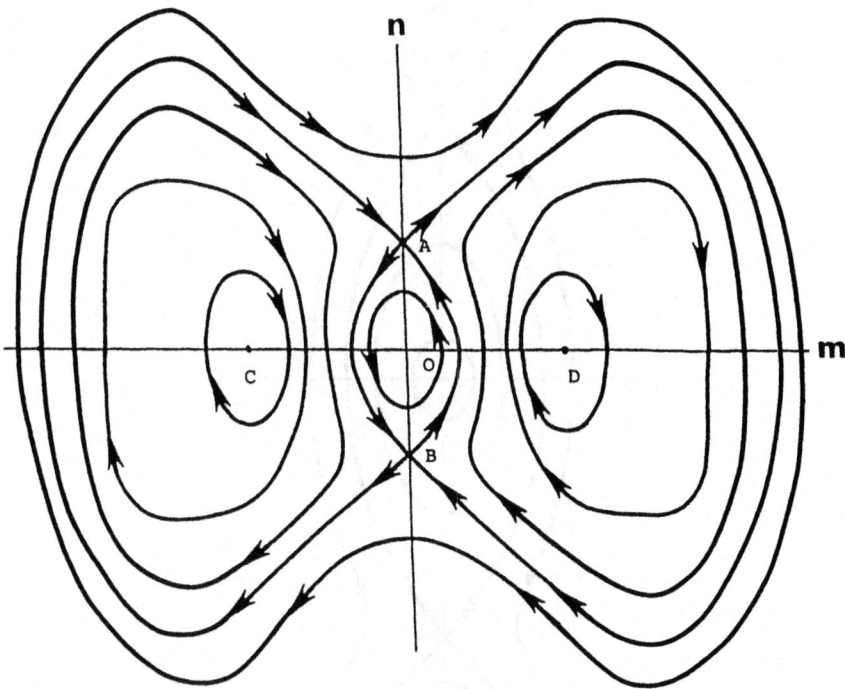

FIGURE 6.13. The WKB phase space for the case in which the bifurcation parameter λ is less than the secondary bifurcation value λ_c, for the case presented in Figure 6.10. $A(0,a)$ and $B(0,-a)$ are saddle points; $C(-a\sqrt{2},0)$, $O(0,0)$, and $D(a\sqrt{2},0)$ are centers.

(6.19) becomes

$$(m^2 + n^2)(m^2 + n^2 - 2a^2) = C'. \tag{6.23}$$

The curves (6.23) in the WKB phase space actually constitute a family of circles, as shown in Figure 6.12. The results demonstrate that wave packet structural vacillations exist on the entire phase space. Except when $m^2 + n^2 = a^2$, they are stable. In general, the curves at the secondary bifurcation point consist of a family of ellipses. It is rather interesting to note that in Figure 6.12, the direction of the evolution of the wave packet inside the circle $m^2 + n^2 = a^2$ is opposite to that outside the circle.

When the topography parameter λ decreases beyond the secondary bifurcation point, Figure 6.13 can be similarly obtained. Now, near the largest spatial scale state and the purely longitudinal spatial scale states, the evolution of a wave packet appears periodic in time, while the purely latitudinal spatial scale states are unstable. Therefore, the largest spatial scale state and the purely longitudinal spatial scale states correspond to centers on the WKB phase space denoted by $O(0,0)$ and $C(-M,0)$ and $D(M,0)$, respectively, where M is defined in (6.7b). However, the purely latitudinal

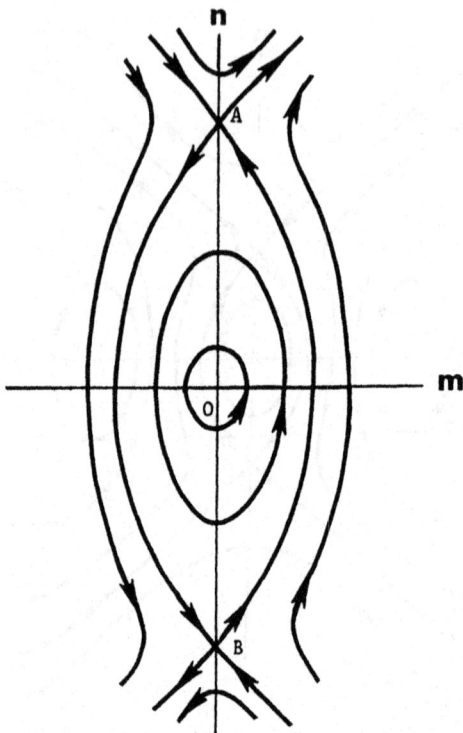

FIGURE 6.14. The WKB phase space for the case in which the topography parameter λ is less than the primary bifurcation value, that is, zero, and when the wave packet is located in a region where the basic flow satisfies $k_1 < 0$ and $k_2 < 0$ on a concave topography. $A(0, N)$ and $B(0, -N)$ are saddle points; $O(0, 0)$ is a center.

spatial scale states correspond to the saddle points $A(0, N)$ and $B(-N, 0)$ in the WKB phase space.

The second example is the case in which the wave packet is located on the right-hand side of a southeasterly jet with a concave topography. The bifurcation diagrams are shown in Figure 6.6. When the bifurcation parameter is less than the primary bifurcation value ($\lambda < 0$), there are three equilibrium states. One is the largest spatial scale state and the other two are the purely latitudinal spatial scale states. The largest spatial scale state is a center $O(0, 0)$, while the two purely latitudinal scale states are two saddle points located at $A(0, N)$ and $B(0, -N)$ in the WKB phase space, where N is determined by (6.7a). Thus we have Figure 6.14 as its configuration. From this figure, it is found that near $O(0, 0)$, the evolution of a wave packet exhibits wave packet structural vacillations. The trajectories connecting $A(0, N)$ and $B(0, -N)$ show two *heteroclinic orbits* of the system, which consist of a homoclinic cycle. Outside the *homoclinic cycle*, the

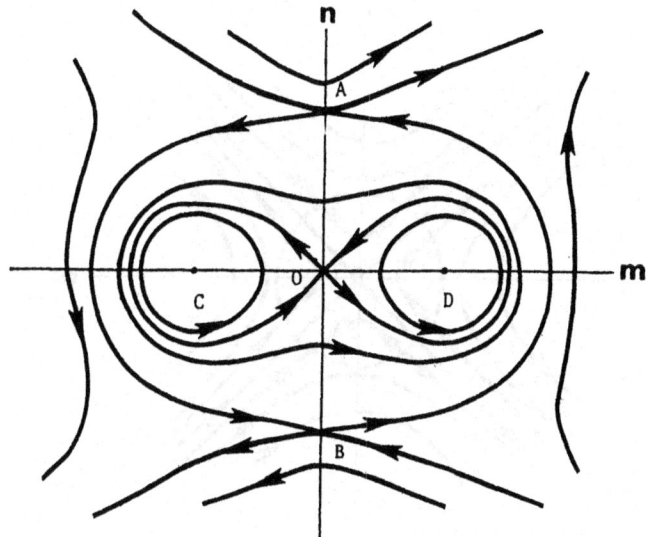

FIGURE 6.15. The WKB phase space for the case in which the bifurcation parameter λ is greater than the primary bifurcation value and less than the secondary bifurcation value, as for the case in Figure 6.10. $A(0, N)$, $O(0, 0)$, and $B(0, -N)$ are saddle points; $C(-M, 0)$ and $D(M, 0)$ are centers.

evolution of a wave packet could not have the packet's structural vacillation properties.

After the bifurcation parameter crosses the primary bifurcation point ($\lambda = \lambda^p = 0$), but before it reaches the secondary bifurcation point ($\lambda = \mu$, when $k_1 = k_2$), there are five equilibrium states in the WKB phase space. The largest spatial scale state becomes unstable, corresponding to a saddle point in the WKB phase space. However, the purely longitudinal spatial scale states become two centers in the WKB phase space, and the purely latitudinal spatial scale states are unstable, becoming the saddle points in the WKB phase space. The WKB trajectories are shown in Figure 6.15. In the figure, $O(0, 0)$, $A(0, N)$, and $B(0, -N)$ are saddle points, whereas $C(-M, 0)$ and $D(M, 0)$ are now centers.

When the topography parameter λ increases to the secondary bifurcation point ($\lambda = \lambda_c$), the WKB trajectories constitute a family of hyperbolic curves. A direct substitution shows that there are the following common integrals:

$$k_1 m^2 - k_2 n^2 = C, \qquad (6.24)$$

where C is a constant of integration. Figure 6.16 shows the configuration in this case, where the arrows show the flow direction of the evolution of a wave packet in the WKB phase space, and $k_1 = k_2$ for simplicity. It is also quite interesting that the points on the circle $m^2 + n^2 = a^2$ in Quadrants

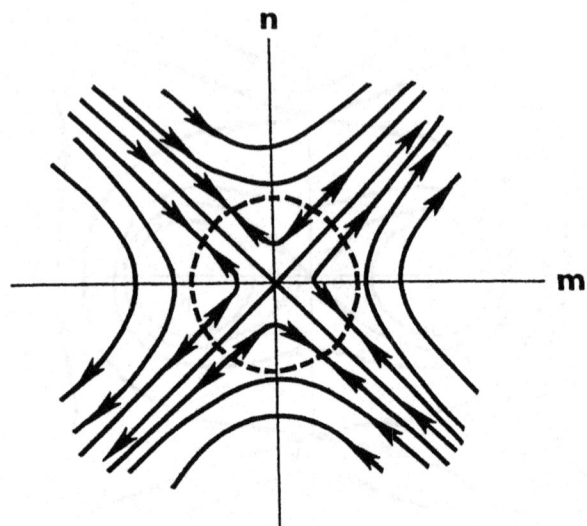

FIGURE 6.16. The WKB phase space for the case in which the bifurcation parameter λ is at the secondary bifurcation point, as for the case in Figure 6.10. $O(0,0)$ is a saddle point. The dashed line corresponds to the circle $m^2 + n^2 = a^2$.

I and III are unstable equilibria, whereas in Quadrants II and IV, the points on the circle are stable equilibria but not asymptotically stable. Here, the equilibria in Quadrants II and IV on the circle mean that the whole spatial scale will eventually have the same total wave number but that the orientation of the wave packet can differ, due to the perturbation. It is the first finding of this kind of stability in real geophysical fluid dynamics.

As the topography parameter λ crosses the secondary bifurcation point, Figure 6.17 can be similarly obtained. Now, the largest spatial scale state is still unstable while the purely longitudinal spatial scale states become unstable, too. The purely latitudinal spatial scale states are neutrally stable. Hence, on the WKB phase space, the largest spatial scale state and the purely longitudinal spatial scale states correspond to saddle points $O(0,0)$, $C(-M,0)$, and $D(M,0)$. The purely latitudinal spatial scale states correspond to centers $A(0,N)$ and $B(0,-N)$ on the WKB phase space.

Furthermore, we are able to obtain the WKB integral in the WKB phase space for the general case, where $k_1 + k_2 \neq 0$. The solutions are

$$k_1^2(m^2 + n^2)^2 + 2k_1\mu(m^2 + n^2) - (k_1 + k_2)m^2 - \frac{\lambda k_1 - \mu k_2}{k_1 + k_2}$$
$$\cdot \ln|(k_1 + k_2)[\mu + k_1(m^2 + n^2)] + \lambda k_1 - \mu k_2| = C, \qquad (6.25)$$

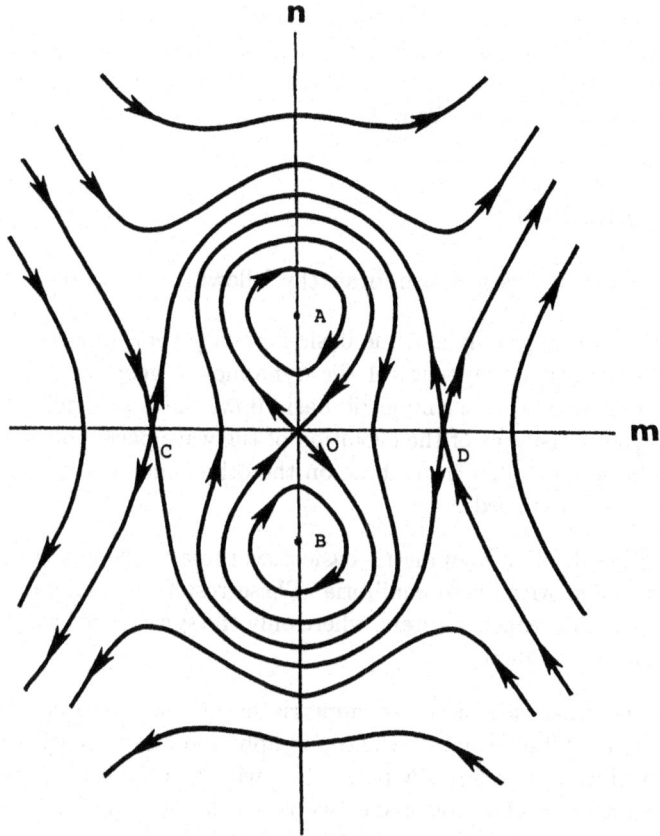

FIGURE 6.17. The WKB phase space for the case in which the bifurcation parameter λ is greater than the secondary bifurcation value, as for the case in Figure 6.10. $A(0, a)$ and $B(0, -a)$ are centers; $C(-a\sqrt{2}, 0)$, $O(0, 0)$, and $D(a\sqrt{2}, 0)$ are saddle points.

where C is a constant of integration to be determined by the initial condition of the wave packet.

From the above analysis and discussion, it is found that on one side of an asymmetric basic flow, there can exist three kinds of wave packet structural vacillations. The first kind of wave vacillation is characterized by its center located at $O(0, 0)$ on the WKB phase space and is called *Vacillation I*. The second kind of vacillation is characterized by its center at $A(0, N)$ or $B(0, -N)$ and is called *Vacillation II*. Vacillation II is restricted to a range of longitudes. The third kind of vacillation is characterized by its center at $C(-M, 0)$ or $D(M, 0)$ and is called *Vacillation III*. Vacillation III is restricted to a range of latitudes. The results demonstrate that on one side of an asymmetric basic flow, the transitions among these three kinds of wave vacillations can occur as the topography parameter is varied.

However, on the other side of the asymmetric basic flow, there are only two kinds of wave vacillations: Vacillation I and Vacillation II or Vacillation III. Transitions can occur only between the two kinds of wave vacillations as the topography parameter changes.

6.6   Conclusions

From the above analysis, we can draw the following conclusions:

1. Both the topography and the basic flow play very important roles in the bifurcation properties of the dynamics of geophysical fluids. In the presence of an asymmetric basic flow, the topological structure and the transitions of the evolution of the wave packet on one side of the basic flow differ from those on the other side, as the topography parameter is varied.

2. On one side of an asymmetric basic flow, there exists only the primary bifurcation with three equilibria. These results are similar to those discussed in earlier chapters, where only the symmetric basic flow was taken into account.

3. On the other side of the asymmetric basic flow, however, there is a cascading bifurcation. As the topography parameter reaches the first critical value, the primary bifurcation with five equilibria occurs. The equilibrium states are, respectively, the largest spatial scale state, two purely longitudinal spatial scale states, and two purely latitudinal spatial scale states. When the topography parameter is increased further to the second critical value, a secondary bifurcation with the mixed spatial scale states appears. The primary bifurcation is a pitchfork bifurcation. The secondary bifurcation, however, is always transcritical.

4. Three kinds of wave packet structural vacillations (oscillations) have been found; they are Vacillation I, Vacillation II, and Vacillation III. The results strongly suggest that the evolution of the wave packet can change its characteristics among the three kinds of wave vacillations from one to another only on one side of an asymmetric basic flow, as the topography parameter is varied. However, on the other side of the basic flow, the transitions can only occur between Vacillation I and Vacillation II or Vacillation III, as the topography parameter changes; this is similar to the previous results or to Yang (1988b), in which only a symmetric basic flow is present.

The bifurcation properties are summarized in Table 6.1 and Table 6.2. The first column of the tables is the conditions of parameters and the second column presents the equilibria. The third and fourth column are type of bifurcation and the simple bifurcation diagrams, respectively.

TABLE 6.1. Summary of bifurcation properties when $\mu k_1 > 0$.

Parameter conditions	Equilibria	Type of Bifurcation		Diagram
		Primary	Secondary	
$\mu > 0$, $k_1 > 0$, $k_2 < 0$	$m = n = 0$; $m^2 = -\lambda/k_2$ $n = 0$	Supercritical	None	
$\mu < 0$, $k_1 < 0$, $k_2 > 0$	$m = n = 0$; $m^2 = -\lambda/k_2$, $n = 0$	Reverse supercritical	None	
$\mu > 0$, $k_1 > 0$, $k_2 > 0$	$m = n = 0$; $m^2 = -\lambda/k_2$ $n = 0$	Subcritical	None	
$\mu < 0$, $k_1 < 0$, $k_2 < 0$	$m = n = 0$; $m^2 = -\lambda/k_2$, $n = 0$	Reverse subcritical	None	

TABLE 6.2. Summary of bifurcation properties when $\mu k_1 < 0$.

Parameter conditions	Equilibria	Type of Bifurcation		Diagram
		Primary	Secondary	
$\mu > 0,$ $k_1 < 0,$ $k_2 < 0$	$m = n = 0;$ $m = 0,$ $n^2 = -\mu/k_1;$	Supercritical	Transcritical	
$\mu > 0,$ $k_1 < 0,$ $k_2 > 0$	$m^2 = -\lambda/k_2,$ $n = 0;$	Subcritical	Transcritical	
$\mu < 0,$ $k_1 > 0,$ $k_2 < 0$	$m^2 = -\lambda/k_2 - n^2,$ $n,$ arbitrary	Reverse subcritical	Transcritical	
$\mu < 0,$ $k_1 > 0,$ $k_2 > 0$		Reverse supercritical	Transcritical	

REFERENCES

Andronov, A.A., Vitt, E.A., and Khaiken, S.E. (1966). *The Theory of Oscillators*. Pergamon Press, Oxford.

Elberry, R. L. (1968). A high-rotating general circulation model experiment with cyclic time changes. Atmospheric Science Paper No. 134, Colorado State University, Fort Collins.

Erneux, T., and Reiss, E. (1983). Splitting of steady multiple eigenvalues may lead to periodic cascading bifurcation. *SIAM J. Appl. Math.* **43**, 613–624.

Fultz, D., Long, R.R., Owens, G.V., Bohan, W., Kaylor, R., and Weil, J. (1959). Studies of thermal convection in a rotating cylinder with some implications for large-scale atmospheric motions. *Meteorol. Monogr.* No. 21.

Gollub, J.P., and Benson, S.V. (1980). Many routes to turbulent convection. *J. Fluid Mech.* **100**, 449–470.

Gruber, A. (1985). The wave number-frequency spectra of the 200 mb wind field in the tropics. *J. Atmos. Sci.* **32**, 1615–1625.

Guckenheimer, J., and Holmes, P. (1983). *Nonlinear Oscillations, Dynamical Systems, and Bifurcations of Vector Fields.* Springer-Verlag, New York.

Hart, J.E. (1984). A laboratory study of baroclinic chaos on the f-plane. *Tellus* **37A**, 286–296.

Hide, R. (1958). An experimental study of thermal convection in a rotating liquid. *Phil. Trans. Roy. Soc. London* **250A**, 441–478.

Krishnamurti, R. (1970). On the transition to turbulent convection. *J. Fluid Mech.* **42**, 295–320.

Krishnamurti, R. (1973). Some further studies on the transition to turbulent convection. *J. Fluid Mech.* **60**, 285–303.

Landau, L.D. (1944). Turbulence. *Dokl. Akad. Nauk. USSR* **44**, 339–340.

Landau, L.D., and Lifshitz, F.M. (1959). *Fluid Mechanics.* Pergamon Press, London.

Landau, L.D., and Lifschitz, F.M. (1987). *Fluid Mechanics,* 2nd ed. Pergamon, Oxford.

Lorenz, E.N. (1963). Deterministic nonperiodic flow. *J. Atmos. Sci.* **20**, 130–141.

McGuirk, J.P., and Reiter, E.R. (1976). A vacillation in atmospheric energy parameters. *J. Atmos. Sci.* **33**, 2079–2093.

Nicolis, G., and Prigogine, I. (1977). *Self-Organization in Nonequilibrium Systems.* John Wiley, New York.

Pedlosky, J., and Frenzen, C. (1980). Chaotic and periodic behavior of finite amplitude baroclinic waves. *J. Atmos. Sci.* **37**, 1177–1196.

Pfeffer, R.L., and Chiang, Y. (1967). Two kinds of vacillation in rotating laboratory experiments. *Mon. Wea. Rev.* **95**, 75–82.

Reiss, E. (1983). Cascading bifurcations. *SIAM J. Appl. Math.* **43**, 57–65.

Sparrow, C. (1982). *The Lorenz Equations: Bifurcation, Chaos, and Strange Attractors.* Springer-Verlag, New York.

Webster, P.J., and Keller, J.L. (1975). Atmospheric variations: Vacillations and index cycles. *J. Atmos. Sci.* **32**, 1283–1300.

Weng, H., Barcilon, A., and Magnan, J. (1986). Transitions between baroclinic flow regimes. *J. Atmos. Sci.* **43**, 1760–1777.

Yang, H. (1988a). Global behavior of the evolution of a Rossby wave packet in barotropic flows on the earth's δ-surface. *J. Atmos. Sci.* **45**, 113–126.

Yang, H. (1988b). Bifurcation properties of the evolution of a Rossby wave packet in barotropic flows on the earth's δ-surface. *J. Atmos. Sci.* **45**, 3667–3683.

Yang, H. (1988c). Secondary bifurcation of the evolution of a Rossby wave packet in barotropic flows on the earth's δ-surface. *J. Atmos. Sci.* **45**, 3684–3699.

7

Evolution of Wave Packets in Stratified Baroclinic Basic Flow

7.1 Introduction

So far, we have discussed only the dynamics of wave packet with the barotropic basic flow in a two-dimensional system. The baroclinicity of the basic flow is of great importance in geophysical fluids. Therefore, in this chapter we extend the previous results (Yang, 1988b,c) into a three-dimensional disturbance system with the baroclinic basic flow. The vorticity equation of motion and Ertel theorem are given in Section 7.2. The model is a three-dimensional stratified disturbance system with barotropic and baroclinic basic flow on the earth's δ-surface. The potential vorticity equation in a stratified geophysical flow is derived in Section 7.3. The equations determining the structure and its change are addressed in Section 7.4. Integral properties of this model are presented in Section 7.5. Sections 7.6 and 7.7 discuss the structures of wave packets and their changes on the zonal and meridional basic flow, respectively.

7.2 The Vorticity Equation and Ertel Theorem

We now try to derive the vorticity equation, since the vorticity is always a very important vector governing motion in geophysical fluids. We first rewrite the momentum eq. (1.5) as follows:

$$\frac{\partial \mathbf{u}}{\partial t} + (2\mathbf{\Omega} + \omega) \times \mathbf{u} = -\frac{1}{\rho}\nabla p + \nabla\left\{\Phi - \frac{|\mathbf{u}|^2}{2}\right\} + \mathbf{F}, \qquad (7.1)$$

where

$$\omega \equiv \nabla \times \mathbf{u} \qquad (7.2)$$

and the vector identity

$$\omega \times \mathbf{u} = (\mathbf{u} \cdot \nabla)\mathbf{u} - \nabla\left(\frac{|\mathbf{u}|^2}{2}\right) \qquad (7.3)$$

has been used.

To derive an equation for ω, we now take the curl of (7.1) and obtain

$$\frac{\partial \omega}{\partial t} + \nabla \times \{(2\Omega + \omega) \times \mathbf{u}\} = \frac{\nabla \rho \times \nabla p}{\rho^2} + \nabla \times \mathbf{F}. \tag{7.4}$$

In order to get a more compact form of vorticity equation, we use another vector identity. For any two vectors $\mathbf{A}$ and $\mathbf{B}$, we have

$$\nabla \times (\mathbf{A} \times \mathbf{B}) = \mathbf{A} \nabla \cdot \mathbf{B} + (\mathbf{B} \cdot \nabla)\mathbf{A} - \mathbf{B} \nabla \cdot \mathbf{A} - (\mathbf{A} \cdot \nabla)\mathbf{B}, \tag{7.5}$$

so that

$$\begin{aligned}
\nabla \times \{(2\Omega + \omega) \times \mathbf{u}\} \\
= (2\Omega + \omega)\nabla \cdot \mathbf{u} + (\mathbf{u} \cdot \nabla)(2\Omega + \omega) - (2\Omega + \omega) \cdot \nabla \mathbf{u}, \tag{7.6}
\end{aligned}$$

since $\omega + 2\Omega$ has zero divergence. Therefore, the vorticity equation can be written in the following compact form:

$$\frac{d\omega}{dt} = \omega_{\mathbf{a}} \cdot \nabla \mathbf{u} - \omega_{\mathbf{a}} \nabla \cdot \mathbf{u} + \frac{\nabla \rho \times \nabla p}{\rho^2} + \nabla \times \mathbf{F}, \tag{7.7}$$

where $\omega_{\mathbf{a}}$ is the absolute vorticity, defined by

$$\omega_{\mathbf{a}} = \omega + 2\Omega. \tag{7.8}$$

From (7.7), it can be seen that the rate of change in the vorticity is due to the sum of the following factors:

(1) $(\nabla \rho \times \nabla p)/\rho^2$: the production of vorticity by baroclinicity;

(2) $\nabla \times \mathbf{F}$: the diffusive effects of friction in the Newtonian flow;

(3) $\omega_{\mathbf{a}} \nabla \cdot \mathbf{u}$: the vortex-tube stretching, which changes the vorticity parallel to the filament by convergence of the filaments; and

(4) $\omega_{\mathbf{a}} \cdot \nabla \mathbf{u}$: the vortex tilting by the variation, along the direction of the filament of the velocity component perpendicular to the filaments of $\omega_{\mathbf{a}}$.

Although the vorticity equation is very useful because it deals directly with the vector character of vorticity, it is more or less a description of how vorticity is changed, not a useful constraint on the change. There does exist an unusually useful and elegant constraint on vorticity, called the *Ertel theorem,* from Ertel's 1942 work. We will discuss the theorem as follows.

Since Ω is a constant vector, the vorticity equation (7.7) can be written as

$$\frac{d\omega_{\mathbf{a}}}{dt} = \omega_{\mathbf{a}} \cdot \nabla \mathbf{u} - \omega_{\mathbf{a}} \nabla \cdot \mathbf{u} + \frac{\nabla \rho \times \nabla p}{\rho^2} + \nabla \times \mathbf{F}. \tag{7.9}$$

The continuity equation (1.3) can be written as

$$\nabla \cdot \mathbf{u} = -\frac{1}{\rho}\frac{d\rho}{dt},$$ (7.10)

so that by eliminating $\nabla \cdot \mathbf{u}$ from (7.9), we obtain

$$\frac{d}{dt}\left(\frac{\omega_{\mathbf{a}}}{\rho}\right) = \left(\frac{\omega_{\mathbf{a}}}{\rho}\cdot\nabla\right)\mathbf{u} + \frac{\nabla\rho\times\nabla p}{\rho^3} + \frac{1}{\rho}\nabla\times\mathbf{F}.$$ (7.11)

Now we consider some scalar fluid property, Θ, which satisfies an equation of the form

$$\frac{d\Theta}{dt} = Q,$$ (7.12)

where Q is an unspecified source of the property Θ.

We then have

$$\frac{\omega_{\mathbf{a}}}{\rho}\cdot\frac{d}{dt}\nabla\Theta = \left(\frac{\omega_{\mathbf{a}}}{\rho}\cdot\nabla\right)\frac{d\Theta}{dt} - \left(\frac{\omega_{\mathbf{a}}}{\rho}\cdot\nabla\mathbf{u}\right)\cdot\nabla\Theta.$$ (7.13)

If the dot product of $\nabla\Theta$ and (7.11) is taken, we obtain

$$\nabla\Theta\cdot\frac{d}{dt}\left(\frac{\omega_{\mathbf{a}}}{\rho}\right) = \left\{\left(\frac{\omega_{\mathbf{a}}}{\rho}\cdot\nabla)\mathbf{u}\right\}\cdot\nabla\Theta + \nabla\Theta\cdot\left\{\frac{\nabla\rho\times\nabla p}{\rho^3}\right\} + \frac{\nabla\Theta}{\rho}\cdot\{\nabla\times\mathbf{F}\},$$ (7.14)

and adding (7.13) and (7.14) together yields

$$\frac{d}{dt}\left\{\frac{\omega_{\mathbf{a}}}{\rho}\cdot\nabla\Theta\right\} = \frac{\omega_{\mathbf{a}}}{\rho}\cdot\nabla Q + \nabla\Theta\cdot\left\{\frac{\nabla\rho\times\nabla p}{\rho^3}\right\} + \frac{\nabla\Theta}{\rho}\cdot\nabla\times\mathbf{F}.$$ (7.15)

Therefore, the Ertel theorem states that if

(1) Θ is a conserved quantity for each fluid element;

(2) the frictional force is negligible, that is, $\mathbf{F} = 0$, and either

(3a) the fluid is barotropic, i.e., $\nabla\rho\times\nabla p = 0$; or

(3b) Θ can be considered a function of p and ρ only,

then the quantity

$$\Pi = \frac{\omega + 2\Omega}{\rho}\cdot\nabla\Theta$$ (7.16)

is conserved by each fluid element, that is,

$$\frac{d\Pi}{dt} = 0.$$ (7.17)

Quantity Π is called the *potential vorticity*.

7.3   The Potential Vorticity Equation

In the inviscid adiabatic fluid, the potential temperature is conserved and is a function of p and ρ if the ideal gas state equation for the atmosphere is applied. The potential temperature θ is defined by

$$\theta = T\left(\frac{p_0}{p}\right)^{R/C_p},\qquad(7.18)$$

where p_0 is a constant reference pressure and R and C_p are the gas constant and specific heat at constant pressure, respectively.

Therefore in the atmosphere, Π can be considered as

$$\Pi = \frac{\omega_\mathbf{a} \cdot \nabla\theta}{\rho},\qquad(7.19)$$

where

$$\omega_\mathbf{a} = \mathbf{k}(f + \zeta) + \mathbf{j}(\eta + f\cot\phi) + \mathbf{i}\xi.\qquad(7.20)$$

The unit vectors $\mathbf{i}$, $\mathbf{j}$, and $\mathbf{k}$ are eastward, northward, and upward, respectively, and ζ, ξ, and η are the vertical and horizontal components of vorticity.

We introduce the nondimensional variables and scalings as follows:

$$(x, y) = L(x', y'),\qquad z = Dz',\qquad(7.21)$$

while the characteristic time is scaled by the advective time, that is,

$$t = \frac{L}{U}t'.\qquad(7.22)$$

For the velocity

$$(u, v) = U(u', v')\qquad(7.23)$$

and

$$w = \frac{D}{L}Uw'.\qquad(7.24)$$

Using these scaling analysis for synoptic disturbance system, we can obtain (Pedlosky, 1987)

$$p = p_s(z) + \rho_s U f_0 L p',\qquad(7.25)$$

$$\rho = \rho_s(z)[1 + \varepsilon F \rho'],\qquad(7.26)$$

and

$$\theta = \theta_s(z)[1 + \varepsilon F \theta'],\qquad(7.27)$$

where ε is the Rossby number defined in (3.13) and F is the Froude number defined in (3.14).

By the use of nondimensional variables and dropping the primes, Π can be expressed as

$$\Pi = \left[\frac{f_0/D}{\rho_s(1+\varepsilon F\rho)}\right]\left\{1+\varepsilon\left(\zeta+\beta_0 y-\frac{\bar{\delta}}{2}y^2\right)\left[\frac{\partial\theta}{\partial z}+\varepsilon F\frac{\partial}{\partial z}(\theta\theta_s)\right]\right.$$
$$\left.+\varepsilon F\left[\varepsilon\zeta\frac{\partial\theta}{\partial x}+\varepsilon\eta\frac{\partial\theta}{\partial y}+\varepsilon\frac{L}{D}\cot\phi_0\frac{\partial\theta}{\partial y}\right]\right\}. \tag{7.28}$$

And the operator d/dt in dimensionless variables is

$$\frac{d}{dt}=\frac{U}{L}\left[\frac{\partial}{\partial t}+\frac{U}{L}\left(u\frac{\partial}{\partial x}+v\frac{\partial}{\partial y}\right)+\frac{W}{D}w\frac{\partial}{\partial z}\right]. \tag{7.29}$$

If only terms of $O(\varepsilon^2,\varepsilon F,\varepsilon\partial\theta_s/\partial z)$ are retained, the Ertel theorem becomes

$$\frac{d_0}{dt}\left\{\zeta+\beta_0 y-\frac{\bar{\delta}}{2}y^2\right\}+\frac{F}{\frac{\rho_0}{\theta_s}\frac{\partial\theta_s}{\partial z}}\frac{d_0}{dt}\frac{\partial\theta_0}{\partial z}+\frac{\rho_s w_1}{\frac{\partial\theta_s}{\partial z}}\frac{\partial}{\partial z}\left(\frac{\frac{\partial\theta_s}{\partial z}}{\rho_s}\right)=0, \tag{7.30}$$

in which the variables have been expanded in the Rossby number series, that is,

$$u=u_0+\varepsilon u_1+\varepsilon^2 u_2+\cdots, \tag{7.31}$$

$$\vdots$$

$$\frac{d_0}{dt}=\frac{\partial}{\partial t}+u_0\frac{\partial}{\partial x}+v_0\frac{\partial}{\partial y}+\cdots. \tag{7.32}$$

The thermodynamics equation can be stated as

$$\frac{d\theta}{dt}=0, \tag{7.33}$$

which, being nondimensionized, becomes

$$\frac{d_0\theta}{dt}+\frac{w}{\varepsilon F\theta_s}\frac{\partial\theta_s}{\partial z}(1+\varepsilon F\theta)=0. \tag{7.34}$$

To the lowest order in ε, eq. (7.34) becomes

$$\frac{d_0\theta_0}{dt}+w_1 S=0, \tag{7.35}$$

where the stratification parameter is defined by

$$S(z)=\frac{1}{F\theta_s}\frac{\partial\theta_s}{\partial z}. \tag{7.36}$$

Using (7.36), we obtain the potential vorticity equation in the stratified fluid, that is,

$$\frac{d_0}{dt}\left[\zeta_0+\beta_0 y-\frac{\bar{\delta}}{2}y^2+\frac{1}{\rho_s}\frac{\partial}{\partial z}\left(\frac{\rho_s}{S}\theta_0\right)\right]=0. \tag{7.37}$$

In the geostrophic approximation,

$$v_0 = \frac{\partial p_0}{\partial x} \tag{7.38}$$

and

$$u_0 = -\frac{\partial p_0}{\partial y}, \tag{7.39}$$

and by the hydrostatic approximation, it can be shown that

$$\theta_0 = \frac{\partial p_0}{\partial z} \tag{7.40}$$

for the problem considered here (Pedlosky, 1987).

The potential vorticity equation now can be written in terms of the stream function $\psi = p_0$:

$$\left(\frac{\partial}{\partial t} - \frac{\partial \psi}{\partial y}\frac{\partial}{\partial x} + \frac{\partial \psi}{\partial x}\frac{\partial}{\partial y}\right)\left\{\frac{\partial^2 \psi}{\partial x^2} + \frac{\partial^2 \psi}{\partial y^2} + \frac{1}{\rho_s}\frac{\partial}{\partial z}\left(\frac{\rho_s}{S}\frac{\partial \psi}{\partial z}\right) + \beta_0 y - \frac{\bar{\delta}}{2}y^2\right\} = 0. \tag{7.41}$$

To simplify the analysis, we consider that S is constant and that

$$-\frac{1}{\rho_s}\frac{\partial \rho_s}{\partial z} = \frac{1}{H} \tag{7.42}$$

is constant. Here H is the density scale height divided by D.

By use of the transformation

$$\psi \rightarrow e^{z/2H}\psi, \tag{7.43}$$

and linearizing the resulting equation of the (7.41), we obtain the perturbation equation

$$\left(\frac{\partial}{\partial t} + U\frac{\partial}{\partial x} + V\frac{\partial}{\partial y}\right)\left(-\frac{\psi}{4H^2 S} + \frac{1}{S}\frac{\partial^2 \psi}{\partial z^2} + \frac{\partial^2 \psi}{\partial y^2} + \frac{\partial^2 \psi}{\partial x^2}\right)$$
$$+ \frac{\partial \psi}{\partial x}\left(\frac{U}{4H^2 S} - \frac{1}{S}\frac{\partial^2 U}{\partial z^2} - \frac{\partial^2 U}{\partial y^2} + \frac{\partial^2 V}{\partial x\partial y} + \beta_0 - \bar{\delta}y\right)$$
$$- \frac{\partial \psi}{\partial y}\left(-\frac{V}{4H^2 S} + \frac{1}{S}\frac{\partial^2 V}{\partial z^2} + \frac{\partial^2 V}{\partial x^2} - \frac{\partial U}{\partial x\partial y}\right) = 0, \tag{7.44}$$

where U and V are eastward and northward components of basic flow, respectively.

Suppose

$$U = U(y, z, t), \qquad V = V(x, z, t); \tag{7.45}$$

eq. (7.44) then becomes

$$\left(\frac{\partial}{\partial t}+U\frac{\partial}{\partial x}+V\frac{\partial}{\partial y}\right)\left(-\frac{\psi}{4H^2S}+\frac{1}{S}\frac{\partial^2\psi}{\partial z^2}+\frac{\partial^2\psi}{\partial y^2}+\frac{\partial^2\psi}{\partial x^2}\right)$$

$$+\frac{\partial\psi}{\partial x}\left(\frac{U}{4H^2S}-\frac{1}{S}\frac{\partial^2 U}{\partial z^2}-\frac{\partial^2 U}{\partial y^2}+\beta_0-\bar{\delta}y\right)$$

$$-\frac{\partial\psi}{\partial y}\left(-\frac{V}{4H^2S}+\frac{1}{S}\frac{\partial^2 V}{\partial z^2}+\frac{\partial^2 V}{\partial x^2}\right)=0. \tag{7.46}$$

7.4　The Equations Governing Amplitude and Structure

The potential vorticity equation (7.46) can be written in the form

$$\left(\frac{\partial}{\partial t}+U\frac{\partial}{\partial x}+V\frac{\partial}{\partial y}\right)\left(-\frac{\psi}{4H^2S}+\frac{1}{S}\frac{\partial^2\psi}{\partial z^2}+\frac{\partial^2\psi}{\partial y^2}+\frac{\partial^2\psi}{\partial x^2}\right)$$

$$+B_1\frac{\partial\psi}{\partial x}-B_2\frac{\partial\psi}{\partial y}=0, \tag{7.47}$$

where

$$B_1=\frac{U}{4H^2S}-\frac{1}{S}\frac{\partial^2 U}{\partial z^2}-\frac{\partial^2 U}{\partial y^2}+\beta_0-\bar{\delta}y, \tag{7.48a}$$

and

$$B_2=-\frac{V}{4H^2S}+\frac{1}{S}\frac{\partial^2 V}{\partial z^2}+\frac{\partial^2 V}{\partial x^2}. \tag{7.48b}$$

Suppose that the basic flow is slowly varying with respect to space and time, as in earlier chapters, and we introduce the slowly varying variables

$$T=\varepsilon t,\qquad X=\varepsilon x,\qquad Y=\varepsilon y,\qquad Z=\varepsilon z \tag{7.49}$$

and

$$\psi=\Psi(X,Y,Z,T)e^{i\theta(X,Y,Z,T)/\varepsilon}, \tag{7.50}$$

where

$$\Psi=\Psi_0(X,Y,Z,T)+\varepsilon\Psi_1(X,Y,Z,T)+\varepsilon^2\Psi_2(X,Y,Z,T)+\cdots. \tag{7.51}$$

Define

$$\sigma=-\frac{\partial\theta}{\partial T},\qquad m=\frac{\partial\theta}{\partial X},\qquad n=\frac{\partial\theta}{\partial Y},\qquad \ell=\frac{\partial\theta}{\partial Z}, \tag{7.52}$$

and we have relations

$$\frac{\partial\sigma}{\partial X}=-\frac{\partial m}{\partial T},\qquad \frac{\partial\sigma}{\partial Y}=-\frac{\partial n}{\partial T},\qquad \frac{\partial\sigma}{\partial Z}=-\frac{\partial\ell}{\partial T} \tag{7.53}$$

and

$$\frac{\partial m}{\partial Y} = \frac{\partial n}{\partial X}, \qquad \frac{\partial m}{\partial Z} = \frac{\partial \ell}{\partial X}, \qquad \frac{\partial n}{\partial Z} = \frac{\partial \ell}{\partial Y}. \tag{7.54}$$

Since B_1 and B_2 are supposed to be slowly varying quantities, by substituting eqs. (7.49) to (7.52) into the potential vorticity equation (7.47), we obtain, at the lowest order, the dispersion relationship

$$\sigma = Um + Vn - \frac{m}{K^2}B_1 + \frac{n}{K^2}B_2, \tag{7.55}$$

where

$$K^2 = m^2 + n^2 + \frac{\ell^2}{S} + \frac{1}{4H^2 S}. \tag{7.56}$$

At the next order, we have the amplitude equation as follows:

$$\left(\frac{\partial}{\partial T} + U\frac{\partial}{\partial X} + V\frac{\partial}{\partial Y}\right)(K^2 \Psi_0)$$

$$- (\sigma - Um - Vn)\left\{2\left(m\frac{\partial}{\partial X} + n\frac{\partial}{\partial Y} + \ell\frac{\partial}{\partial Z}\right)\Psi_0\right.$$

$$\left. + \Psi_0\left(\frac{\partial m}{\partial X} + \frac{\partial n}{\partial Y} + \frac{\partial \ell}{\partial Z}\right)\right\} - B_1\frac{\partial \Psi_0}{\partial X} + B_2\frac{\partial \Psi_0}{\partial Y} = 0. \tag{7.57}$$

From the dispersion relation (7.56), we have

$$C_X = \frac{\sigma}{m} = U + V\frac{n}{m} - \frac{B_1}{K^2} + \frac{n}{mK^2}B_2, \tag{7.58}$$

$$C_Y = \frac{\sigma}{n} = U\frac{m}{n} + V - \frac{m}{nK^2}B_1 + \frac{B_2}{K^2}, \tag{7.59}$$

and

$$C_Z = \frac{\sigma}{\ell} = U\frac{m}{\ell} + V\frac{n}{\ell} - \frac{m}{\ell K^2}B_1 + \frac{n}{\ell K^2}B_2. \tag{7.60}$$

Here, C_X, C_Y, and C_Z are the three components of the phase velocity. The group velocity is found to be

$$C_{gX} = \frac{\partial \sigma}{\partial m} = U - K^{-4}\{B_1 K^2 - 2m(B_1 m - B_2 n)\}, \tag{7.61}$$

$$C_{gY} = \frac{\partial \sigma}{\partial n} = V + K^{-4}\{B_2 K^2 + 2n(B_1 m - B_2 n)\}, \tag{7.62}$$

and

$$C_{gZ} = \frac{\partial \sigma}{\partial \ell} = \frac{2\ell}{SK^4}(B_1 m - B_2 n). \tag{7.63}$$

By the use of the theory developed in Chapter 2, we can derive the following equations:

$$\frac{D_g \sigma}{DT} = -\left\{m\frac{\partial U}{\partial T} + n\frac{\partial V}{\partial T} - \frac{m}{K^2}\frac{\partial B_1}{\partial T} + \frac{n}{K^2}\frac{\partial B_2}{\partial T}\right\}. \tag{7.64}$$

$$\frac{D_g m}{DT} = -\left\{ m\frac{\partial U}{\partial X} + n\frac{\partial V}{\partial X} - \frac{m}{K^2}\frac{\partial B_1}{\partial X} + \frac{n}{K^2}\frac{\partial B_2}{\partial X} \right\}, \tag{7.65}$$

$$\frac{D_g n}{DT} = -\left\{ m\frac{\partial U}{\partial Y} + n\frac{\partial V}{\partial Y} - \frac{m}{K^2}\frac{\partial B_1}{\partial Y} + \frac{n}{K^2}\frac{\partial B_2}{\partial Y}, \tag{7.66}$$

and

$$\frac{D_g \ell}{DT} = -\left\{ m\frac{\partial U}{Z} + n\frac{\partial V}{\partial Z} - \frac{m}{K^2}\frac{\partial B_1}{\partial Z} + \frac{n}{K^2}\frac{\partial B_2}{\partial Z} \right\}. \tag{7.67}$$

In the case that

$$U = U(Y, Z, T), \qquad V = V(X, Z, T), \tag{7.68}$$

the equations become

$$\frac{D_g m}{DT} = -\frac{n}{K^2}\left\{ \frac{\partial V}{\partial X}\left(m^2 + n^2 + \frac{\ell^2}{S} \right) + \frac{\partial A_m}{\partial X} \right\}, \tag{7.69}$$

$$\frac{D_g n}{DT} = -\frac{m}{K^2}\left\{ \frac{\partial U}{\partial Y}\left(m^2 + n^2 + \frac{\ell^2}{S} \right) + \frac{\partial A_n}{\partial Y} + \delta_0 \right\}, \tag{7.70}$$

and

$$\frac{D_g \ell}{DT} = -\frac{n}{K^2}\left\{ \frac{\partial V}{\partial Z}\left(m^2 + n^2 + \frac{\ell^2}{S} \right) + \frac{\partial A_m}{\partial Z} \right\}$$
$$\qquad - \frac{m}{K^2}\left\{ \frac{\partial U}{\partial Z}\left(m^2 + n^2 + \frac{\ell^2}{S} \right) + \frac{\partial A_n}{\partial Z} \right\}, \tag{7.71}$$

where A_m and A_n are the strength of the jet-like meridional and the zonal basic flow, respectively. They are defined by

$$A_m = \frac{\partial^2 V}{S\partial z^2} + \frac{\partial^2 V}{\partial x^2} \tag{7.72}$$

and

$$A_n = \frac{\partial^2 U}{S\partial z^2} + \frac{\partial^2 U}{\partial y^2}. \tag{7.73}$$

The δ_0 is related to $\bar{\delta}$ by (3.40).

The equations governing the changes in the three-dimensional total wave number and the meridional and vertical tilts of trough surface $(-n/m, -\ell/m)$ can be easily obtained from (7.69) to (7.71). The equation governing the three-dimensional total wave number is

$$\frac{D_g}{DT}(m^2 + n^2 + \ell^2)^{1/2} = -\frac{1}{(m^2 + n^2 + \ell^2)^{1/2}} \cdot$$
$$\cdot \left\{ \frac{mn}{K^2}\left[\frac{\partial V}{\partial X}\left(m^2 + n^2 + \frac{\ell^2}{S} \right) + \frac{\partial A_m}{\partial X} \right] \right.$$

$$+ \frac{mn}{K^2}\left[\frac{\partial U}{\partial Y}\left(m^2 + n^2 + \frac{\ell^2}{S}\right) + \frac{\partial A_n}{\partial Y} + \delta_0\right]$$

$$+ \frac{n\ell}{K^2}\left[\frac{\partial V}{\partial Z}\left(m^2 + n^2 + \frac{\ell^2}{S}\right) + \frac{\partial A_m}{\partial Z}\right]$$

$$\left.+ \frac{m\ell}{K^2}\left[\frac{\partial U}{\partial Z}\left(m^2 + n^2 + \frac{\ell^2}{S}\right) + \frac{\partial A_n}{\partial Z}\right]\right\}. \quad (7.74)$$

The equation of meridional tilt of trough surface is

$$\frac{D_g}{DT}\left(-\frac{n}{m}\right) = \frac{1}{K^2}\left\{\frac{\partial U}{\partial Y}\left(m^2 + n^2 + \frac{\ell^2}{S}\right) + \frac{\partial A_n}{\partial Y} + \delta_0\right\}$$

$$- \frac{n^2}{m^2 K^2}\left\{\frac{\partial V}{\partial X}\left(m^2 + n^2 + \frac{\ell^2}{S}\right) + \frac{\partial A_m}{\partial X}\right\}. \quad (7.75)$$

The equation for the vertical tilt of trough surface is

$$\frac{D_g}{DT}\left(-\frac{\ell}{m}\right) = \frac{1}{K^2}\left\{\frac{\partial U}{\partial Z}\left(m^2 + n^2 + \frac{\ell^2}{S}\right) + \frac{\partial A_n}{\partial Z}\right\}$$

$$+ \frac{n}{mK^2}\left\{\frac{\partial V}{\partial Z}\left(m^2 + n^2 + \frac{\ell^2}{S}\right) + \frac{\partial A_m}{\partial Z}\right\}$$

$$- \frac{n\ell}{m^2 K^2}\left\{\frac{\partial V}{\partial X}\left(m^2 + n^2 + \frac{\ell^2}{S}\right) + \frac{\partial A_m}{\partial X}\right\}. \quad (7.76)$$

We now have obtained all the equations governing the evolution of the wave packet in the three-dimensional stratified baroclinic basic flow.

7.5   Integral Properties and Instability Theorems

In the absence of meridional basic flow, the amplitude equation (7.57) becomes

$$\left(\frac{\partial}{\partial T} + U\frac{\partial}{\partial X}\right)(K^2\Psi_0) - (\sigma - Um)\left\{2\left(m\frac{\partial}{\partial X} + n\frac{\partial}{\partial Y} + \ell\frac{\partial}{\partial Z}\right)\Psi_0\right.$$

$$\left.+ \Psi_0\left(\frac{\partial m}{\partial X} + \frac{\partial n}{\partial Y} + \frac{\partial \ell}{\partial Z}\right)\right\} - B_1\frac{\partial \Psi_0}{\partial X} = 0. \quad (7.77)$$

We now let

$$\Psi_0 = |\Psi_0(X, Y, Z, T)|e^{i\alpha(X,Y,Z,T)}. \quad (7.78)$$

Substituting (7.78) into (7.77), we obtain

$$\frac{D_g\alpha}{DT} = 0 \quad (7.79)$$

and

$$\frac{D_g}{DT}(K^2|\Psi_0|) + \frac{1}{2}K^2|\Psi_0|\nabla \cdot \mathbf{C_g} - \frac{K^2|\Psi_0|}{2B_1}\mathbf{C_g}\cdot\nabla B_1 = 0 \quad (7.80)$$

or

$$\frac{D_g}{DT}\left(\frac{K^2}{2}|\Psi_0|^2\right) + \frac{1}{2}K^2|\Psi_0|^2\nabla\cdot\mathbf{C_g} = m\left(n\frac{\partial U}{\partial Y} + \ell\frac{\partial U}{\partial Z}\right)|\Psi_0|^2. \quad (7.81)$$

Integrating (7.81) over the whole region in the study (W), we obtain

$$\frac{\partial}{\partial T}\iiint_{(W)}\frac{K^2}{2}|\Psi_0|^2 dXdYdZ$$

$$= \iiint_{(W)} m\left(n\frac{\partial U}{\partial Y} + \ell\frac{\partial U}{\partial Z}\right)|\Psi_0|^2 dXdYdZ. \quad (7.82)$$

Similarly, in virtue of (7.74) and (7.81), the following integral can be derived:

$$\iiint_{(W)} B_1^{-1}\frac{\partial}{\partial T}(K^4|\Psi_0|^2)dXdYdZ = 0. \quad (7.83)$$

Using (7.82) and (7.83), we can further obtain

$$\iiint\left\{\frac{\partial}{\partial T}(K^2|\Psi_0|^2) + \frac{U_r - U}{B_1}\frac{\partial}{\partial T}(K^4|\Psi_0|^2)\right\}dXdYdZ = 0, \quad (7.84)$$

where U_r is the reference zonal basic flow. These three integral equations are only the extension of the previous ones in barotropic basic flow, as discussed in Section 3.4. The integral properties share similar meanings as do those in barotropic basic flow in Section 3.4. Equation (7.82) states the conservation of the wave packet's energy. Equation (7.83) and eq. (7.84) are the analogues of conservation of wave-action and energy-modified enstrophy.

The total wave packet energy is given by

$$E = \iiint_{(W)}\frac{1}{2}K^2|\Psi_0|^2 dXdYdZ. \quad (7.85)$$

If we define $\mathbf{K}^0$ as the unit wave vector, that is,

$$\mathbf{K}^0 \equiv \frac{1}{(m^2+n^2+\ell^2)^{1/2}}(\mathbf{i}m + \mathbf{j}n + \mathbf{k}\ell), \quad (7.86)$$

then eq. (7.82) can be written as

$$\frac{\partial}{\partial T}\iiint_{(W)} K^2|\Psi_0|^2 dXdYdZ$$

$$= \iiint_{(W)} (m^2+n^2+\ell^2)m\frac{\partial U}{\partial\mathbf{K}^0}|\Psi_0|^2 dXdYdZ. \quad (7.87)$$

In the same sense as in Section 3.4, the three instability theorems can be extended to the stratified baroclinic basic flow, as stated in the following:

Theorem I. If the basic flow is positioned with respect to the wave packet, such as $(\partial U/\partial K^0) > 0$, then the wave packet is unstable; otherwise, it is stable.

Theorem II. The necessary condition for the instability of wave packet is that there is at least one point at which $B_1 = 0$, that is,

$$B_1 \equiv \frac{U}{4H^2S} - \frac{\partial^2 U}{S\partial z^2} - \frac{\partial^2 U}{\partial y^2} + \beta_0 - \delta_0 Y = 0, \ \text{at} \ Y = Y_j, Z = Z_j. \quad (7.88)$$

Theorem III. The necessary condition for the instability of wave packet is that at least some where there is

$$\frac{U_r - U}{B_1} < 0. \qquad (7.89)$$

The proofs of the integral properties and the theorems are similar to those given in Section 3.4.

7.6   The Structures of the Developing and Decaying Wave Packet

As discussed in Section 3.4, in general, an unstable wave packet will not always be unstable due to its evolution with time. Since the wave packet is changing with the slowly varying variables, the wave packet is said to be *developing* or *decaying* if its total energy is increased or decreased with time (Zeng, 1982, 1983a,b). Since m is always taken to be positive, the direction of development of the wave packet is completely determined by the gradient of the zonal basic flow with respect to the wave packet from eq. (7.87). The wave packet is developing when $(\partial U/\partial K^0) > 0$ and decaying when $(\partial U/\partial K^0) < 0$.

It can be shown that the whole spatial scale of the wave packet represented by the inverse of the total wave number will increase when the wave packet is developing and decrease when the wave packet is decaying, as discussed in Section 3.4. Therefore, the concept of developing and decaying here can be expressed either in term of total energy or the whole spatial scale of wave packet and is consistent with common sense.

7.6.1   IN THE PURELY BAROTROPIC ZONAL BASIC FLOW

In the purely barotropic zonal basic flow, we have

$$\frac{\partial U}{\partial Z} = 0. \qquad (7.90)$$

Suppose that there is a purely barotropic, jet-like zonal basic flow with a maximum at Y_0, as shown in Figure 7.1; we then have $(\partial U/\partial Y) > 0$,

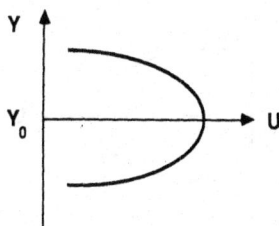

FIGURE 7.1. An ideal barotropic westerly jet.

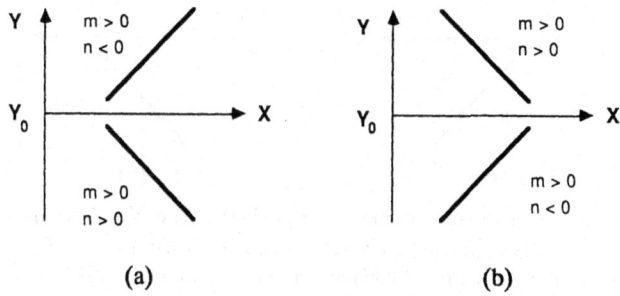

FIGURE 7.2. The structure of the wave packet in the (X, Y) plane when there is only a purely barotropic basic westerly flow, as shown in Figure 7.1. (a) Barotropically developing wave packet and (b) barotropically decaying wave packet.

when $Y < Y_0$, and $(\partial U/\partial Y) < 0$ when $Y > Y_0$. For the barotropically developing wave packet, we can conclude from (7.87) that $n(\partial U/\partial K^0) > 0$. Therefore, the meridional tilt on the (X, Y) plane should be westward, that is, $n > 0$ for $Y < Y_0$, and eastward, that is, $n < 0$ for $Y > Y_0$, as shown in Figure 7.2a, where the heavy solid line is the trough line or ridge line of the barotropically developing wave packet in the (X, Y) plane. Similarly, we obtain the structure of barotropically decaying wave packet, as shown in Figure 7.2b. From these figures, it is found that if the wave packet is occupied over the whole area of the barotropic, jetlike zonal basic flow, then the wave packet with the same meridional tilt either eastward or westward will have an unstable structure. That is to say, on one side of the zonal jet the structure of the wave packet is barotropically developing, while on the other side of the jet the wave packet is barotropically decaying when the wave packet is either eastward tilted or westward tilted with latitude throughout the whole study area.

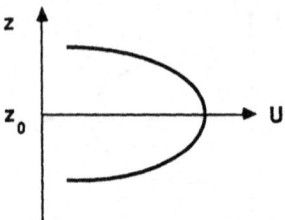

FIGURE 7.3. An ideal baroclinic westerly jet.

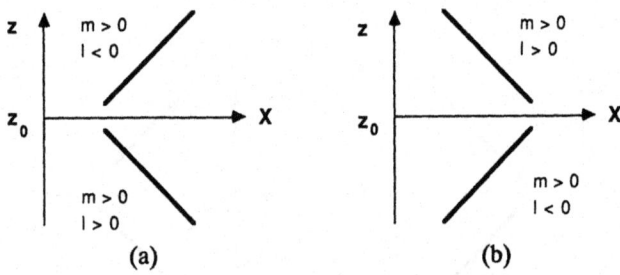

(a)   (b)

FIGURE 7.4. The structure of the wave packet in the (X, Z) plane when there is only a purely baroclinic westerly jet, as shown in Figure 7.3. The heavy solid lines are vertical trough lines (surfaces in three dimensions) of the wave packet. (a) Baroclinically developing wave packet and (b) baroclinically decaying wave packet.

7.6.2 IN THE PURELY BAROCLINIC ZONAL BASIC FLOW

In a purely baroclinic basic flow, we have

$$\frac{\partial U}{\partial Y} = 0. \tag{7.91}$$

The results are similar to those in the purely barotropic zonal basic flow. Again, suppose the purely baroclinic basic flow with the maximum at $Z = Z_0$ is in the presence of the wave packet, as shown in Figure 7.3, with similar arguments; we then conclude that the baroclinically developing wave packet is tilted eastward, that is, $\ell < 0$, for $Z > Z_0$ and westward, that is, $\ell > 0$, for $Z < Z_0$, as given in Figure 7.4a. The baroclinically decaying wave packet has the structure that it is tilted westward with an increase of altitude for $Z > Z_0$ and eastward for $Z < Z_0$, as illustrated in Figure 7.4b. Similarly, if the wave packet is occupied over the whole area of the baroclinic basic flow with the same vertical tilt either westward or eastward, the wave packet is unstable. That is, the upper layer of the wave packet is baroclinically developing (or decaying), while the low layer of the wave packet is baroclinically decaying (or developing).

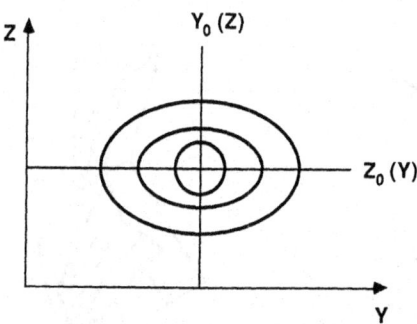

FIGURE 7.5. An ideal westerly jet with both barotropicity and baroclinicity. Its center is located at (Y_0, Z_0).

7.6.3 IN THE BAROTROPIC AND BAROCLINIC ZONAL BASIC FLOW

In the presence of both barotropicity and baroclinicity in the basic flow, the trough or ridge will be a three-dimensional surface in the (X, Y, Z) space. If there is a barotropic and baroclinic westerly jet, such as the one in Figure 7.5, the jet reaches its maximum at (Y_0, Z_0) and is symmetrical with respect to the point (Y_0, Z_0).

The structure of the barotropically and baroclinically developing wave packet can be found in the same manner. It can be concluded that the structure of the barotropically and baroclinically developing wave packet has the characteristic such that its trough or ridge surface (1) is tilted eastward with altitude and latitude in the range where $Y > Y_0$ and $Z > Z_0$, (2) is tilted westward with altitude and latitude in the range where $Y < Y_0$ and $Z < Z_0$, (3) is tilted westward with latitude and eastward with altitude in the range $Y < Y_0$ and $Z > Z_0$, and (4) is tilted eastward with latitude and westward with altitude in the range $Y > Y_0$ and $Z < Z_0$. The structure of the barotropically and baroclinically developing wave packet is shown in Figure 7.6a.

Figure 7.6b illustrates the structure of the barotropically and baroclinically decaying wave packet. From this figure, one finds that the trough surface of the barotropically and baroclinically decaying wave packet (1) is tilted westward with latitude and altitude in the range where $Y > Y_0$ and $Z > Z_0$, (2) is tilted eastward with latitude and altitude in the range $Y < Y_0$ and $Z < Z_0$, (3) is tilted eastward with latitude and westward with altitude in the range $Y < Y_0$ and $Z > Z_0$, and (4) is tilted westward with latitude and eastward with altitude in the range $Y > Y_0$ and $Z < Z_0$.

The structure of the barotropically decaying, but baroclinically developing wave packet is characterized such that its trough surface (1) is tilted westward with latitude and eastward with altitude in the range $Y > Y_0$ and $Z > Z_0$, (2) is tilted eastward with latitude and westward with alti-

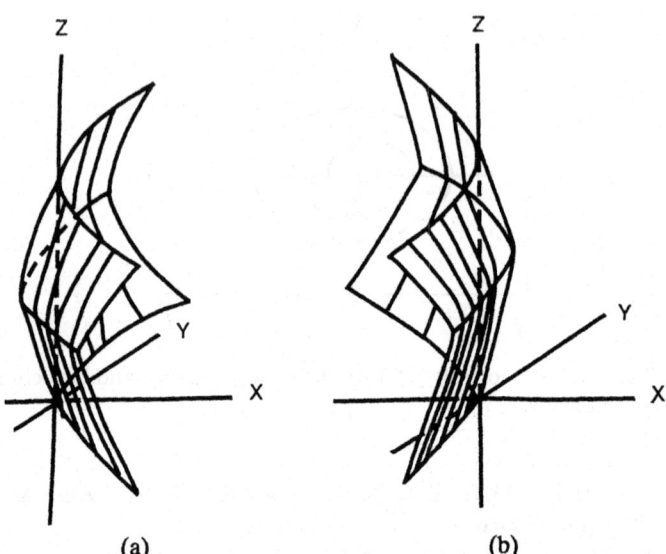

(a)   (b)

FIGURE 7.6. The structure of the wave packet when both barotropicity and baro-clinicity are present in the basic flow. (a) The barotropically and baroclinically developing wave packet and (b) the barotropically and baroclinically decaying wave packet.

tude in the range $Y < Y_0$ and $Z < Z_0$, (3) is tilted westward with latitude and altitude in the range $Y > Y_0$ and $Z < Z_0$, and (4) is tilted eastward with latitude and altitude in the range $Y < Y_0$ and $Z > Z_0$, as shown in Figure 7.7a. The structure of a barotropically developing but baroclinically decaying wave packet is given in Figure 7.7b. This kind of wave packet has the feature such that its trough surface (1) is tilted eastward with latitude and westward with altitude in the range $Y > Y_0$ and $Z > Z_0$, (2) is tilted westward with latitude and altitude in the range $Y < Y_0$ and $Z > Z_0$, (3) is tilted westward with latitude and eastward with altitude in the range $Y < Y_0$ and $Z < Z_0$, and (4) is tilted westward with latitude and eastward with altitude in the range $Y > Y_0$ and $Z < Z_0$.

Similarly, we can obtain the structure in the purely meridional basic flow, for the barotropically and baroclinically developing and decaying, barotropically developing but baroclinically decaying, and barotropically decaying but baroclinically developing wave packets, which are similar to those in the zonal basic flow. The structure of the wave packets in the presence of the asymmetrical basic flow with both horizontal and vertical shears is more complicated. However, the structures can, in principle, be determined by the use of the same arguments.

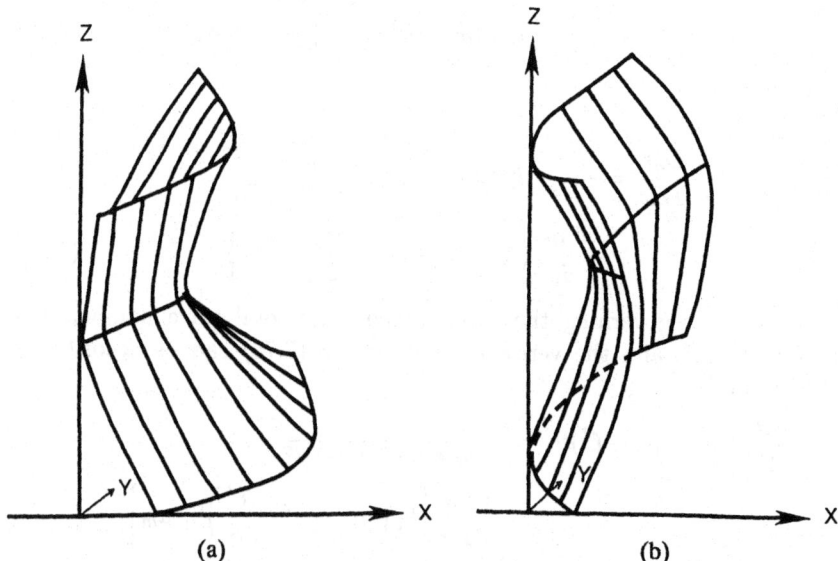

FIGURE 7.7. The same as Figure 7.6, except (a) is the barotropically decaying, but baroclinically developing wave packet; and (b) is the barotropically developing, but baroclinically decaying wave packet.

7.7 Structure Changes and Bifurcations

Although in the preceding section we have discussed the structure of developing and decaying wave packet, the structure is also changing with the evolution of the wave packet. As mentioned before, a developing wave packet will not always remain developing. For instance, it is possible that a baroclinically developing wave packet will become baroclinically decaying, and a barotropically decaying wave packet will become barotropically developing, or vice versa. A barotropically decaying but baroclinically developing wave packet also might become barotropically developing but baroclinically decaying. It is the change in the structure of the disturbance system that is of great importance in the daily forcast and study of geophysical fluid dynamics. Only when we know the change in the structure can we have a better understanding of the nature of geophysical fluids.

As mentioned in the preceding section, the increase or decrease in the energy of the wave packet (developing or decaying wave packet) can be described by the change in the three-dimensional total wave number of the wave packet. Therefore, the change in structure also shows the change in wave packet energy.

For convenience, the equations governing the change in the wave number (7.69) to (7.71) can be rewritten as follows:

$$\frac{D_g m}{DT} = -\frac{n}{K^2}\left\{ k_1\left(m^2 + n^2 + \frac{\ell^2}{S}\right) + a_m\right\}, \tag{7.92}$$

$$\frac{D_g n}{DT} = -\frac{m}{K^2}\left\{k_2\left(m^2 + n^2 + \frac{\ell^2}{S}\right) + a_n + \delta_0\right\}, \qquad (7.93)$$

and

$$\frac{D_g \ell}{DT} = -\frac{n}{K^2}\left\{k_2\left(m^2 + n^2 + \frac{\ell^2}{S}\right) + b_m\right\}$$
$$-\frac{m}{K^2}\left\{k_4\left(m^2 + n^2 + \frac{\ell^2}{S}\right) + b_n\right\}. \qquad (7.94)$$

The equations governing the three-dimensional total wave number, the meridional tilt, and the vertical tilt (7.74) to (7.76) can be accordingly rewritten as

$$\frac{D_g}{DT}(m^2 + n^2 + \ell^2)^{1/2} = -\frac{1}{(m^2 + n^2 + \ell^2)^{1/2}} \cdot$$
$$\cdot\left\{\frac{mn}{K^2}\left[k_1\left(m^2 + n^2 + \frac{\ell^2}{S}\right) + a_m\right]\right.$$
$$+ \frac{mn}{K^2}\left[k_2\left(m^2 + n^2 + \frac{\ell^2}{S}\right) + a_n + \delta_0\right]$$
$$+ \frac{n\ell}{K^2}\left[k_3\left(m^2 + n^2 + \frac{\ell^2}{S}\right) + b_m\right]$$
$$\left.+ \frac{m\ell}{K^2}\left[k_4\left(m^2 + n^2 + \frac{\ell^2}{S}\right) + b_n\right]\right\}, \qquad (7.95)$$

$$\frac{D_g}{DT}\left(-\frac{n}{m}\right) = \frac{1}{K^2}\left\{k_2\left(m^2 + n^2 + \frac{\ell^2}{S}\right) + a_n + \delta_0\right\}$$
$$-\frac{n^2}{m^2 K^2}\left\{k_1\left(m^2 + n^2 + \frac{\ell^2}{S}\right) + a_m\right\}, \qquad (7.96)$$

and

$$\frac{D_g}{DT}\left(-\frac{\ell}{m}\right) = \frac{1}{K^2}\left\{k_4\left(m^2 + n^2 + \frac{\ell^2}{S}\right) + b_n\right\}$$
$$+ \frac{n}{m K^2}\left\{k_3\left(m^2 + n^2 + \frac{\ell^2}{S}\right) + b_m\right\}$$
$$-\frac{n^2}{m^2 K^2}\left\{k_1\left(m^2 + n^2 + \frac{\ell^2}{S}\right) + a_m\right\}, \qquad (7.97)$$

where

$$k_1 = \frac{\partial V}{\partial X}, \qquad k_2 = \frac{\partial U}{\partial Y}, \qquad k_3 = \frac{\partial V}{\partial Z}, \qquad k_4 = \frac{\partial U}{\partial Z} \qquad (7.98)$$

and

$$a_m = \frac{\partial A_m}{\partial X}, \qquad a_n = \frac{\partial A_n}{\partial Y}, \qquad b_m = \frac{\partial A_m}{\partial Z}, \qquad b_n = \frac{\partial A_n}{\partial Z}. \qquad (7.99)$$

These are parameters of the dynamic system. Here k_1 and k_2 are linear barotropic shears of the meridional basic flow in the X-direction and the zonal basic flow in the Y-direction, respectively, and k_3 and k_4 are, respectively, the linear baroclinic shears of the meridional and zonal basic flow in the vertical direction; a_m is the change in the strength of the jetlike meridional basic flow in the X-direction, and a_n is the change in the strength of the jetlike zonal basic flow in the Y-direction; b_m and b_n represent vertical changes in strength of the meridional and the zonal basic flow, respectively. These parameters are generally considered to be independent, though this might not be true in reality.

7.7.1   IN THE PURELY BAROTROPIC BASIC FLOW

In the absence of baroclinicity in the basic flow, the basic flow is purely barotropic. The governing equations (7.74) to (7.76) can be written as

$$\frac{D_g m}{DT} = -\frac{n}{K^2}\left\{k_1\left(m^2 + n^2 + \frac{\ell^2}{S}\right) + a_m\right\}, \tag{7.100}$$

$$\frac{D_g n}{DT} = -\frac{m}{K^2}\left\{k_2\left(m^2 + n^2 + \frac{\ell^2}{S}\right) + a_n + \delta_0\right\}, \tag{7.101}$$

and

$$\frac{D_g \ell}{DT} = 0. \tag{7.102}$$

Equations (7.74) to (7.76), governing the three-dimensional total wave number and the tilts of the wave packet, become

$$\frac{D_g}{DT}(m^2 + n^2 + \ell^2)^{1/2} = -\frac{1}{(m^2 + n^2 + \ell^2)^{1/2}} \cdot$$
$$\cdot \left\{\frac{mn}{K^2}\left[k_1\left(m^2 + n^2 + \frac{\ell^2}{S}\right) + a_m\right]\right.$$
$$\left. + \frac{mn}{K^2}\left[k_2\left(m^2 + n^2 + \frac{\ell^2}{S}\right) + a_n + \delta_0\right]\right\}, \tag{7.103}$$

$$\frac{D_g}{DT}\left(-\frac{n}{m}\right) = \frac{1}{K^2}\left\{k_2\left(m^2 + n^2 + \frac{\ell^2}{S}\right) + a_n + \delta_0\right\}$$
$$- \frac{n^2}{m^2 K^2}\left\{k_1\left(m^2 + n^2 + \frac{\ell^2}{S}\right) + a_m\right\}, \tag{7.104}$$

and

$$\frac{D_g}{DT}\left(-\frac{\ell}{m}\right) = \frac{m\ell}{m^2 K^2}\left\{k_1\left(m^2 + n^2 + \frac{\ell^2}{S}\right) + a_m\right\}. \tag{7.105}$$

From the governing equations, we see that the vertical spatial scale of the wave packet does not change with time. However, the total energy or the total spatial scale of wave packet varies by the change in the wave numbers in the X-direction and the Y-direction. The meridional tilt and the vertical tilt of the wave packet will also change. The structural change in this purely barotropic basic flow is similar to the one we discussed before. However, now the wave packet is three dimensional and all the terms have been retained this time. When we compare the equations with those in earlier chapters, we find that now the two-dimensional total wave number is replaced by the three-dimensional total wave number, and there are two additional terms in the present form. Moreover, a vertical tilt is also present. If there is only a linear shear in the basic flow, the equations are reduced to the earlier ones.

When we compare the dynamic system with that in earlier chapters, for example, eqs. (4.10) to (4.11), we find that the two dynamic systems are similar when the topography is that $k_0 = 0$, for instance, symmetric. Nevertheless, the topography parameters are replaced by the parameters of the strength of shear of the barotropic basic flow. Moreover, the dynamic system is three dimensional. Despite the difference, using the same method, it can be shown the system possesses similar bifurcation behavior except that now the bifurcation parameter is the strength of the shear of basic flow, which is more relevant to the situation in geophysical fluids than the topography is. Table 7.1 and Table 7.2 summarize the main results of the bifurcation properties in stratified purely barotropic basic flows when the basic flow satisfies the condition that $k_1(k_1\ell^2/S + a_m) > 0$ and $k_1(k_1\ell^2/S + a_m) < 0$, respectively. In these tables, the far left column is the condition under which the bifurcation diagram in the far right column occurs. The second column from the left shows the equilibrium states. The third column from the left indicates the type of primary bifurcation. And the far right column shows the bifurcation diagram. We note that the secondary bifurcation exhibited in Table 7.2 is always a transcritical bifurcation.

In Table 7.1, the conditions I, II, III, and IV, respectively, correspond to the following:

I: $k_1\ell^2/S + a_m > 0$, $k_1 > 0$, $k_2 < 0$;

II: $k_1\ell^2/S + a_m < 0$, $k_1 < 0$, $k_2 > 0$;

III: $k_1\ell^2/S + a_m > 0$, $k_1 < 0$, $k_2 > 0$; and

IV: $k_1\ell^2/S + a_m < 0$, $k_1 < 0$, $k_2 < 0$.

TABLE 7.1. Summary of bifurcation properties in stratified purely barotropic basic flows when $k_1(k_1\ell^2/S + a_m) > 0$.

I		supercritical	
II	$m = n = 0$; $m^2 = -\frac{k_1}{k_2}\frac{\ell^2}{S}$, $-\frac{\lambda + \delta_0}{k_2}$, $n = 0$.	reverse supercritical	
III		subcritical	
IV		reverse subcritical	

In Table 7.2, the conditions I, II, III, and IV, respectively, correspond to the following:

I: $k_1\ell^2/S + a_m > 0$, $k_1 < 0$, $k_2 < 0$;

II: $k_1\ell^2/S + a_m > 0$, $k_1 < 0$, $k_2 > 0$;

TABLE 7.2. Summary of bifurcation properties in stratified purely barotropic basic flows when $k_1(k_1\ell^2/S + a_m) < 0$.

I		supercritical	
II	$m = n = 0;$ $m = 0,$ $n^2 = -\frac{\ell^2}{S}$ $-\frac{a_m}{k_1};$	subcritical	
III	$m^2 = -n^2$ $-\frac{\ell^2}{S}$ $-\frac{\lambda + \delta_0}{k_2},$	reverse subcritical	
IV	n arbitrary.	reverse supercritical	

III: $k_1\ell^2/S + a_m < 0$, $k_1 > 0$, $k_2 < 0$; and

IV: $k_1\ell^2/S + a_m < 0$, $k_1 > 0$, $k_2 > 0$.

The bifurcation parameter now is the change in the strength of jetlike zonal basic flow, that is,

$$\lambda = a_n, \tag{7.106}$$

where we still use λ as the bifurcation parameter. The primary bifurcation point is

$$\lambda^p = -\delta_0 - k_2 \frac{\ell^2}{S}. \tag{7.107}$$

The secondary bifurcation point is

$$\lambda^s = \lambda_c = \frac{a_m}{k_1} k_2 - \delta_0. \tag{7.108}$$

Compared with the results in Chapter 6, we find that the primary bifurcation point and the secondary bifurcation point have shifted, due to the definition of the bifurcation parameter.

It can be seen from the tables that the stratification of the basic state will alter the bifurcation properties. For instance, consider the basic flow satisfying $\lambda \geq 0$. If $k_2 > 0$, the primary bifurcation can occur only when the basic flow is unstably stratified. Similarly, if $k_2 < 0$, then the primary bifurcation can occur only when the basic flow is stably stratified. Moreover, if $a_m = 0$, then only the secondary bifurcation can occur when the basic flow is unstably stratified, though the critical value of the secondary bifurcation does not depend on the stratification parameter. Hereafter, the discussions on unstably stratified basic flow may be purely mathematic arguments.

7.7.2 IN THE PURELY BAROCLINIC BASIC FLOW

When there is only the purely baroclinic basic flow, eqs. (7.92) to (7.97) become

$$\frac{D_g m}{DT} = 0, \tag{7.109}$$

$$\frac{D_g n}{DT} = -\frac{m}{K^2} \delta_0, \tag{7.110}$$

$$\frac{D_g \ell}{DT} = -\frac{n}{K^2} \left\{ k_3 \left(m^2 + n^2 + \frac{\ell^2}{S} \right) + b_m \right\}$$
$$- \frac{m}{K^2} \left\{ k_4 \left(m^2 + n^2 + \frac{\ell^2}{S} \right) + b_n \right\}, \tag{7.111}$$

$$\frac{D_g}{DT} (m^2 + n^2 + \ell^2)^{1/2} = -\frac{1}{(m^2 + n^2 + \ell^2)^{1/2}} \cdot$$
$$\cdot \left\{ \frac{mn}{K^2} \delta_0 + \frac{n\ell}{K^2} \left[k_3 \left(m^2 + n^2 + \frac{\ell^2}{S} \right) \right. \right.$$
$$\left. \left. + b_m \right] + \frac{m\ell}{K^2} \left[k_4 \left(m^2 + n^2 + \frac{\ell^2}{S} \right) + b_n \right] \right\}, \tag{7.112}$$

$$\frac{D_g}{DT} \left(-\frac{n}{m} \right) = \frac{\delta_0}{K^2}, \tag{7.113}$$

and

$$\frac{D_g}{DT}\left(-\frac{\ell}{m}\right) = \frac{n}{mK^2}\left\{k_3\left(m^2 + n^2 + \frac{\ell^2}{S}\right) + b_m\right\}$$
$$+ \frac{1}{K^2}\left\{k_4\left(m^2 + n^2 + \frac{\ell^2}{S}\right) + b_n\right\}. \qquad (7.114)$$

From these equations, it is seen that the wave number in the X-direction will remain the same with time, while the wave number in the Y-direction will change due to the δ-effect, discussed in Chapter 3 or in Yang (1987). On the β-plane, the wave number in the Y-direction will also remain the same. The vertical wave number will change due to the baroclinicity of the basic flow. Therefore, we conclude that the baroclinicity of the basic flow will only alter the vertical spatial scale of the wave packet and has no effect on the horizontal spatial scale, which is just an example of the application of one of the theorems presented in Section 2.4.

We now consider a simple example of linear zonal basic flow. For the linear shear of purely baroclinic zonal basic flow, eq. (7.111) becomes

$$\frac{D_g\ell}{DT} = -\frac{m^2 + n^2 + \ell^2/S}{K^2}mk_4. \qquad (7.115)$$

We can conclude from this equation, that, in a stably stratified purely baroclinic basic flow, the vertical spatial scale of the wave packet that is tilted westward with altitude increases when it is located in the lower layer where $k_4 > 0$ and decreases when it is located in the upper layer where $k_4 < 0$, regardless of the meridional tilt of the wave packet. The wave packet's vertical spatial scale will decrease (increase) on the lower (upper) level when it is tilted eastward with altitude. Let us now look at the change in the vertical tilt due to the baroclinicity of the stably stratified linear zonal basic flow. Equation (7.114) reads

$$\frac{D_g}{DT}\left(-\frac{\ell}{m}\right) = \frac{m^2 + n^2 + \frac{\ell^2}{S}}{K^2}k_4. \qquad (7.116)$$

Equation (7.116) tells us that in the upper layer of the westerly jet, the vertical tilt of the wave packet will be tilted clockwise on the (X, Z) plane, while in the lower layer the tilt will move counterclockwise. Figure 7.8 illustrates the structural change of the three-dimensional spatial scale of a wave packet, due to the baroclinicity of the stably stratified basic flow with linear shear, when it is located in the lower layer. The thin line corresponds to an ideal wave packet at $T = 0$, and the heavy line corresponds to the situation at $T > 0$. It shows that the longitudinal spatial scale remains the same, while the vertical and latitudinal spatial scales increase with time. The vertical tilt of the trough surface in the (X, Z) plane will be tilted clockwise, which is hard to present in a figure.

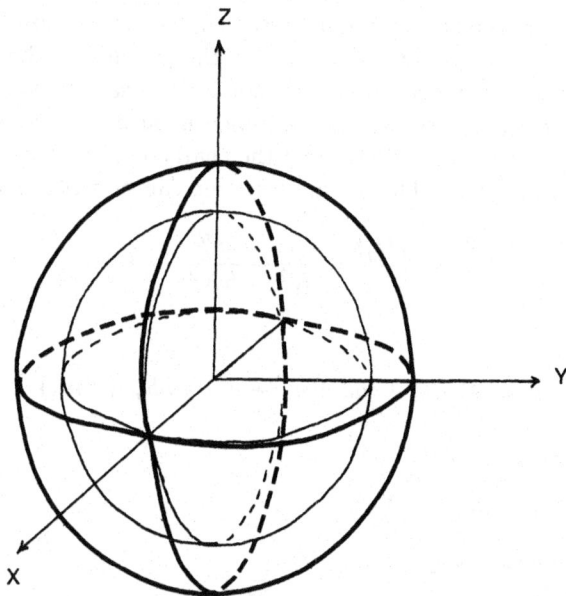

FIGURE 7.8. The three-dimensional spatial scale change of the wave packet in the basic flow with both barotropicity and baroclinicity. The thin lines at $T = 0$, the heavy lines at $T > 0$.

It can be further concluded that in the upper (lower) layer of an unstably stratified purely baroclinic basic flow, the vertical spatial scale of the wave packet tilting westward (eastward) with altitude will decrease when $m^2 + n^2 > -\ell^2/S$ and will increase when $m^2 + n^2 < -\ell^2/S$. However, in the lower (upper) layer of an unstably stratified purely baroclinic basic flow, the vertical spatial scale of a westward (eastward) tilting wave packet will increase when $m^2 + n^2 > -\ell^2/S$ and decrease when $m^2 + n^2 < -\ell^2/S$.

From the above analysis, it is found that in a purely baroclinic linear zonal basic flow, the structure and how it changes a wave packet are independent of the meridional tilt of the wave packet. But they are dependent on the vertical tilt of the wave packet. However, if the basic flow is not zonal, the results are different. For example, we now consider the case in which only a linear purely meridional basic flow is present. Equation (7.111) can be written as

$$\frac{D_g \ell}{DT} = -\frac{m^2 + n^2 + \frac{\ell^2}{S}}{K^2} n k_3. \tag{7.117}$$

From this equation, it is shown that the meridional tilt of a wave packet with latitude will lead to different conclusions if the wave packet is tilted eastward with latitude, though similar conclusions still hold when the wave packet is tilted westward with latitude.

We now discuss the bifurcation properties in a stratified purely baroclinic basic flow. First we consider the case on the β-plane of the earth. The governing equations on the earth's β-plane are the same as eqs. (7.109) to (7.111), with $\delta = 0$. Now it is a one-dimensional dynamic system, since neither m and n vary with time along the characteristic curve, that is, the group velocity direction. The dynamic system can be written as

$$\frac{D_g \ell}{DT} = \frac{\lambda}{K^2} - \frac{a\ell^2}{SK^2},$$

(7.118)

where

$$\lambda = -\{(nk_3 + mk_4)(m^2 + n^2) + nb_m + mb_n\}$$

(7.119a)

and

$$a = nk_3 + mk_4.$$

(7.119b)

The equilibrium states are

$$\ell = \pm\sqrt{\frac{S\lambda}{a}}.$$

(7.120)

The positive state corresponds to the equilibrium vertical spatial scale of the wave packet tilting westward with altitude and the negative state corresponds to the equilibrium vertical spatial scale of the wave packet tilting eastward with altitude.

It is already found that there is a bifurcation at $\lambda = 0$, if $\lambda S/a > 0$, provided $a \neq 0$. Taking λ as a bifurcation parameter, we obtain bifurcation diagrams for the vertical spatial scale of a wave packet as shown in Figure 7.9a and Figure 7.9b in the stratified purely baroclinic basic flow satisfying the conditions $aS > 0$ and $aS < 0$, respectively. This shows that the bifurcation point is a regular turning point. Hence, the bifurcation is a turning point bifurcation, or a saddle-node bifurcation, as discussed in Section 5.2.3.

For example, if we consider the basic flow is a zonal flow, we have

$$a = mk_4$$

(7.121a)

and

$$\lambda = -\{mk_4(m^2 + n^2) + nb_m\}.$$

(7.121b)

The equilibrium states of the wave packet are the same as (7.120). Then in the upper layer where $k_4 < 0$, the bifurcation diagram of the vertical spatial scale of a wave packet has the characteristic shown in Figure 7.9a and Figure 7.9b, respectively, when the wave packet is located in an unstably stratified purely baroclinic basic flow and in a stably stratified purely baroclinic basic flow. However, in the lower layer, where $k_4 > 0$, the bifurcation diagram

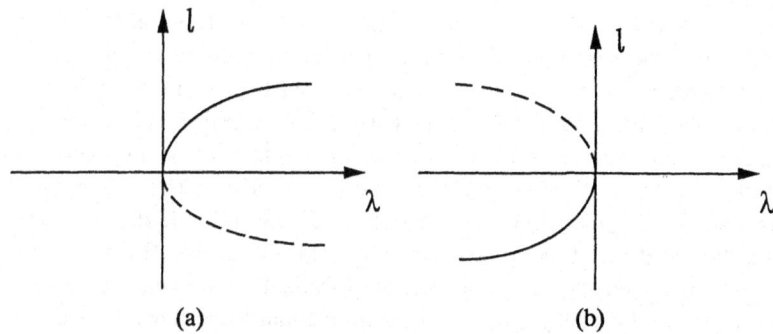

FIGURE 7.9. Turning point bifurcations when (a) $aS > 0$ and (b) $aS < 0$.

for the vertical spatial scale of a wave packet will have the characteristic shown in Figure 7.9a and Figure 7.9b, respectively, when the wave packet is located in a stably stratified purely baroclinic basic flow and in an unstably stratified purely baroclinic basic flow.

Since m is supposed to be positive, the bifurcation diagram in Figure 7.9a demonstrates that a wave packet will be bifurcated into an equilibrium state with a finite vertical spatial scale when it is tilted eastward, while the bifurcation diagram in Figure 7.9b demonstrates that a wave packet will be bifurcated into an equilibrium state with a finite vertical spatial scale when it is tilted westward. The wave packet, hence, will have a fixed eastward tilt or a fixed westward tilt. From these analyses, therefore, we can conclude that in a stably stratified purely baroclinic zonal basic flow, a wave packet will be bifurcated into an equilibrium state with a finite vertical spatial scale and a fixed westward tilt in the upper layer. In the lower layer, however, a wave packet in a stably stratified purely baroclinic zonal basic flow will be bifurcated into an equilibrium state with a finite vertical spatial scale and a fixed eastward tilt. In an unstably stratified purely baroclinic zonal basic flow, a wave packet will be bifurcated into an equilibrium state with a finite vertical spatial scale and a fixed eastward tilt and a fixed westward tilt in the upper and lower layers, respectively.

If we consider a linear purely baroclinic basic flow, however, the bifurcation will only occur in an unstably stratified basic flow. The horizontal spatial scale of the wave packet or the production of the horizontal spatial scale and the stratification parameter can be taken as the varying bifurcation parameter. Similar bifurcation properties can be obtained for a wave packet in both the upper layer and the lower layer.

The implications of the above results for geophysical fluids are very interesting and useful. Now if the stable equilibrium state of the vertical spatial scale of wave packet is interpreted as the vertical localization of disturbance systems, the results strongly suggest that in the lower layer of a baroclinic zonal flow, for example, below the tropopause, the disturbance systems tilting westward with altitude would evolve into a stable equilibrium state with a definite vertical spatial scale in the stably stratified troposphere. The westward tilting disturbance systems in the upper layer would not have the stable equilibrium state with a definite vertical spatial scale. Therefore, all disturbances with different scales will be attracted into a stable equilibrium state with the same definite vertical spatial scale. Since this stable vertical spatial scale only depends on the nature of the media and does not depend on the nature of the individual disturbance, vertical localization of the disturbance system will occur. However, if the disturbances are eastward tilted, vertical localization will not occur in the stably stratified lower layer of the baroclinic zonal basic flow. This may be one of the main reasons why westward tilting disturbance systems in the lower layer are usually more locally characterized than are those in the upper layer. The vertical spatial scales of the disturbance systems are smaller than those in the upper layer. The eastward tilting disturbance systems in the stably stratified lower layer are not persistent. It also may be one of the reasons why disturbance systems tilting westward with altitude are more intense than those in systems in the upper layer. From (7.120), we can also conclude that the equilibrium vertical spatial scale of wave packet is proportional to the inverse square root of the stratification S. The more stably stratified the baroclinic basic flow is, the smaller the equilibrium vertical spatial scale will be, or vice versa. This conclusion agrees with our observation and common sense.

When the change in the strength of basic flow is considered to be the bifurcation parameter, the following equilibrium states can be found on the δ-surface:

$$m = n = 0, \qquad \ell \text{ arbitrary}; \qquad\qquad (7.122)$$

$$m = 0, \qquad n^2 = -\left(\frac{\ell^2}{S} + \frac{b_n}{k_3}\right), \qquad \ell \text{ arbitrary}. \qquad (7.123)$$

The bifurcated states exist only when

$$\frac{b_n}{k_3} < -\frac{\ell^2}{S}. \qquad\qquad (7.124)$$

Bifurcation properties can be considered in the same manner as before. One finds that the equilibrium states are independent of k_4 and b_m because of the presence of the δ-effect.

In the case in which there is only the linear shear in the baroclinic flow, there appears one family of equilibrium states, that is,

$$m = n = 0, \qquad \ell \text{ arbitrary}, \qquad\qquad (7.125)$$

on the δ-surface. On the β-plane, there are equilibrium states when the meridional tilt has the value

$$-\frac{n}{m} = \frac{k_4}{k_3}. \qquad\qquad (7.126)$$

However, if the strength of the baroclinic basic flow changes, the possible equilibrium state is the state when the basic flow satisfies the condition

$$\frac{b_m}{k_3} = \frac{b_n}{k_4}. \tag{7.127}$$

In this case, the equilibrium state in the WKB phase space is an elliptic sphere when the basic flow is stably stratified and a hyperbola when the basic flow is unstably stratified on the earth's β-plane, such as

$$m^2 + n^2 + \frac{\ell^2}{S} = -\frac{b_m}{k_3}. \tag{7.128}$$

As for the β-plane, we can discuss the stabilities of these equilibrium states and their bifurcations on the δ-surface.

7.7.3 IN THE BAROTROPIC AND BAROCLINIC BASIC FLOW

When both the barotropicity and baroclinicity of basic flow are taken into account, the problem becomes more complicated. However, in principle, we can study the structures and their wave packet changes in the same manner. They are combinations of the barotropicity and baroclinicity of the basic flow as discussed in the preceding subsections.

From the governing equations, we find when both barotropicity and baroclinicity are present in the basic flow, there are following possible equilibrium states:

(1) The purely vertical spatial scale state that

$$m = n = 0, \qquad \ell \text{ arbitrary}. \tag{7.129}$$

(2) The latitudinal spatial scale states that

$$m = 0, \qquad n \neq 0$$

and

$$n = \pm\sqrt{-\frac{a_m}{k_1} - \frac{\ell^2}{S}}, \qquad \ell \text{ arbitrary}, \tag{7.130}$$

when

$$\frac{a_m}{k_1} = \frac{b_m}{k_3} < -\frac{\ell^2}{S}. \tag{7.131}$$

(3) The longitudinal spatial scale states that

$$m \neq 0, \qquad n = 0$$

and

$$m = \pm\sqrt{-\frac{a_n + \delta_0}{k_2} - \frac{\ell^2}{S}}, \qquad \ell \text{ arbitrary}, \tag{7.132}$$

when

$$\frac{a_n + \delta_0}{k_2} = \frac{b_n}{k_4} < -\frac{\ell^2}{S}. \tag{7.133}$$

(4) The three-dimensional spatial scale states that

$$m^2 + n^2 + \frac{\ell^2}{S} = -\frac{a_m}{k_1}, \tag{7.134}$$

when

$$\frac{a_m}{k_1} = \frac{a_n + \delta_0}{k_2} = \frac{b_m}{k_3} = \frac{b_n}{k_4} < -\frac{\ell^2}{S}. \tag{7.135}$$

When there is only a zonal basic flow, then $V = 0$, and the equations read

$$\frac{D_g m}{DT} = 0, \tag{7.136}$$

$$\frac{D_g n}{DT} = -\frac{m}{K^2}\left\{k_2\left(m^2 + n^2 + \frac{\ell^2}{S}\right) + a_n + \delta_0\right\}, \tag{7.137}$$

and

$$\frac{D_g \ell}{DT} = -\frac{m}{K^2}\left\{k_4\left(m^2 + n^2 + \frac{\ell^2}{S}\right) + b_n\right\}. \tag{7.138}$$

The equilibrium states are

$$m = 0, \quad n, \quad \ell \text{ arbitrary} \tag{7.139}$$

and

$$m^2 + n^2 + \frac{\ell^2}{S} = -\frac{a_n + \delta_0}{k_2}, \tag{7.140}$$

when

$$\frac{a_n + \delta_0}{k_2} = \frac{b_n}{k_4} < -\frac{\ell^2}{S}. \tag{7.141}$$

When there is only the meridional basic flow, the equations become

$$\frac{D_g m}{DT} = -\frac{n}{K^2}\left\{k_1\left(m^2 + n^2 + \frac{\ell^2}{S}\right) + a_m\right\}, \tag{7.142}$$

$$\frac{D_g n}{DT} = -\frac{m}{K^2}\delta_0, \tag{7.143}$$

and

$$\frac{D_g \ell}{DT} = -\frac{n}{K^2}\left\{k_3\left(m^2 + n^2 + \frac{\ell^2}{S}\right) + b_m\right\}. \tag{7.144}$$

The equilibrium states are found to be

$$m = n = 0, \quad \ell \text{ arbitrary} \tag{7.145}$$

and

$$m = 0, \qquad n^2 + \frac{\ell^2}{S} = -\frac{a_m}{k_1}, \qquad (7.146)$$

when

$$\frac{a_m}{k_1} = \frac{b_m}{k_3} < -\frac{\ell^2}{S}. \qquad (7.147)$$

We could, in principle, investigate the stabilities and their bifurcations using the same method as before. However, we do not present them here. Some detailed results will be published in another form. It should be noted that there is a difference between the consideration here and those in earlier chapters. Firstly, we have to consider the baroclinicity and the vertical structure of the wave packet. Secondly, the stratification will play a role in the study under consideration. For the stably stratified basic flow, $S > 0$, whereas for the unstratified basic flow, $S < 0$. The bifurcation properties, however, can be studied similarly for both stably and unstably stratified basic flows. The stratification has to be taken into account when the bifurcation behavior is studied. Furthermore, if we allow stratification to vary with height, which is the case in real geophysical fluids, then the dynamic system of the structure of the wave packet will be different. It will be very interesting to see what role variation in stratification plays in the bifurcation behavior of the present dynamic system.

7.8 Closure

There are many unsolved problems in this work that could be worked on by people interested in geophysical fluid dynamics or applied mathematics. For example, what is the global behavior of wave packet evolution in the phase space in the presence of horizontal *and* vertical shears in basic flow? The physical significance and relevance of phenomena in geophysical fluids obviously need further work. The validity of the method employed here requires justification, both mathematically and physically. That investigation, in general, will bring out the interesting question about the validity of the approximation theory of partial differential equations in mathematics.

However, from the mathematical point of view, the dynamic behavior of the three nonlinear equations derived here (7.92) to (7.94) itself is an interesting topic. Many questions need to be answered. For instance, Is it possible to have any periodic solutions for this dynamic system? From intuition and our results, the answer may be "yes." And then, under what conditions do such periodic solutions exist? The periodic solution corresponds to the wave packet structural vacillation discussed in earlier chapters and by Yang (1988a). We see that this dynamic system (7.92) to (7.94) is three dimensional and highly nonlinear. From the study of nonlinear dynamical system, we know that in the three-dimensional nonlinear dynamical system chaotic behavior might exist (e.g., Guckenheimer and Holmes, 1983). How does such a nonlinear system behave? Are there any chaotic motions? Are there any strange attractors? Answering these questions will increase our knowledge of nonlinear dynamic systems and of the behavior of geophysical fluids as well.

Our dynamic system was derived from the partial differential equation. Therefore, further investigation should provide more insight into the understanding of such a partial differential equation and its approximations. The relationship between the partial differential equation and the nonlinear dynamic system is also interesting.

In conclusion, the analysis presented here is just the beginning. Further investigations are needed. The problems raised above might be of great importance. However, as we pointed out in our earlier studies, although the method is effective and powerful, we should be extremely careful if we want to apply the results directly to a real physical problem, since after all, the theory is an approximation theory.

REFERENCES

Ertel, H. (1942). Ein neuer hydrodynamischer Wirbesatz. *Meteorolol. Z.* **59**, 277–281.

Guckenheimer, J., and Holmes, P. (1983). *Nonlinear Oscillations, Dynamical Systems, and Bifurcations of Vector Fields.* Springer-Verlag, New York.

Pedlosky, J. (1987). *Geophysical Fluid Dynamics,* 2nd ed. Springer-Verlag, New York.

Yang, H. (1987). Evolution of a Rossby wave packet in barotropic flows with asymmetric basic current, topography and δ-effect. *J. Atmos. Sci.* **44**, 2267–2276.

Yang, H. (1988a). Global behavior of the evolution of a Rossby wave packet in barotropic flows on the earth's δ-surface. *J. Atmos. Sci.* **45**, 113–126.

Yang, H. (1988b). Bifurcation properties of the evolution of a Rossby wave packet in barotropic flows on the earth's δ-surface. *J. Atmos. Sci.* **45**, 3667–3683.

Yang, H. (1988c). Secondary bifurcation of the evolution of a Rossby wave packet in barotropic flows on the earth's δ-surface. *J. Atmos. Sci.* **45**, 3684–3699.

Zeng, Q.-C. (1982). On the evolution and interaction of disturbances and zonal flow in rotating barotropic atmosphere. *J. Meteorol. Soc. Japan, Ser. II* **60**, 24–31.

Zeng, Q.-C. (1983a). The development characteristics of quasi-geostrophic baroclinic disturbances. *Tellus* **35**, 337–347.

Zeng, Q.-C. (1983b). The evolution of a Rossby-wave packet in a three-dimensional baroclinic atmosphere. *J. Atmos. Sci.* **40**, 73–84.

8

Wave Packets and Teleconnections

8.1 Introduction

In this chapter, we discuss another aspect of the wave packet theory, wave packet propagation. Earlier chapters mainly focused on the structure and structural change of wave packets, without explicit discussions of the propagation property of wave packets, which is possible only when the structure independence theorem discussed in Chapter 2 holds. However, when the structure independence theorem is not valid, we have to take the propagation property of the wave packet into account, as discussed in Chapter 2. Moreover, the propagation property of wave packet is also of great importance. As shown in Chapter 2, the wave packet is always propagated along the group velocity, that is, the wave packet velocity. Many modern ideas on wave propagation were originated by Rayleigh, including the distinction between the phase velocity and the group velocity; this appears very early in Rayleigh's work (1877). Since then, there have been numerous studies on wave packet propagation associated with the group velocity, including several monographs (Brillouin, 1946, 1953, 1960; Tolstoy, 1973). The study of wave packet propagation was started as soon as the concept of wave group or wave packet was introduced in wave mechanics (Brillouin, 1946).

In geophysical fluid dynamics, the zonal group velocity of planetary waves was first discussed by Rossby (1945, 1949), who demonstrated that the amplitude of nondivergent planetary waves propagates eastward. Subsequently, Yeh (1949) showed that the group velocity of divergent Rossby waves with large wave lengths is westward relative to the basic flow. Using an influence function, Charney (1949) showed that a numerical prediction at some point is influenced by initial values over some limited region, due to the maximum and minimum values of group velocities.

The vertical propagation of planetary waves was first considered by Charney and Drazin (1961). From their theory, it was found that the energy of stationary planetary waves is trapped in easterlies or strong westerlies and that only large-scale wave can propagage upward into weak westerlies. Dickinson (1968, 1969) found that vertically propagating planetary waves also attenuate with height through Newtonian cooling. Matsno (1970) studied the waves propagating in winter in the Northern Hemisphere. The vertical-zonal propagation of equatorial, mixed Rossby–gravity wave packets forced by a localized tropospheric heat source has been shown by Holton (1972). Hayashi (1981), using a complex Fourier analysis, studied the vertical–zonal propagation of a stationary wave packet and found that, in the troposphere, the wave packet of the time mean geopotential height, consisting of wave numbers 1 through 3, attains its major and minor maxima in the Pacific and Atlantic, respectively.

The zonal–meridional ray paths, that is, the characteristic paths, of barotropic and baroclinic Rossby waves have been discussed by Longuet-Higgins (1964a,b, 1965) and by Schopf et al. (1981), respectively. The numerical demonstration of the zonal–meridional propagation of a barotropic Rossby wave packet was made by Hoskins et al. (1977). Tung and Lindzen (1979a,b) proposed that blocking in the atmosphere could be interpreted in terms of the linear resonance of stationary, vertically, forced propagating planetary waves.

Smith (1971) showed that the energy of a topographic Rossby wave packet tends to be refracted away for oceanic ridges or the equator, absorbed by shore-lines, and trapped along escarpments, by using the wave packet theory, and also found that the wave packet is at least qualitatively accurate, which confirmed the results of Longuet-Higgins (1968) and Rhines (1969). Wave packet propagation has been extensively studied in order to describe the refraction of surface gravity waves in harbors, for example, and readily lends itself to numerical computation (Skovgaard et al., 1975, 1976). Numerical studies of wave packet propagation can be easily found in literature, for example, Vichnevetsky (1981, 1984a,b, 1987) and Trefethen (1982). More references can be found therein.

However, in the present chapter, in Sections 8.3 and 8.4, respectively, we discuss the wave packet propagation on a sphere in zonal flow and the asymmetric basic flow as it was discussed by Hoskins and Karoly (Hoskins and Karoly, 1981; Karoly, 1983). We begin with discussion of teleconnections and go on to the stationary forcing wave packet and teleconnections, as described by Yang and Yang (1988, 1990).

8.2 Teleconnections and the Stationary Forcing Wave Packet

Significant simultaneous correlations between temporal fluctuations in meteorological parameters at widely separated points on the earth are commonly called *teleconnections*. These teleconnections provide evidence that the behavior of the planetary waves is transient. There is abundant evidence that such correlations, particularly in fluctuations with time scales of a week or longer, exist. Five patterns of such teleconnections have been identified by Wallace and Gutzler (1981), who used the National Meteorological Center monthly mean winter data. They are (1) the eastern Atlantic

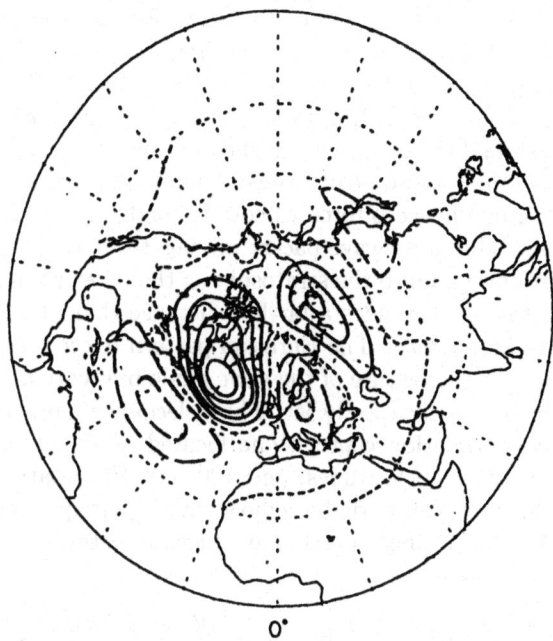

FIGURE 8.1. The western Atlantic pattern at 500 mb. After Hoskins and Karoly (1981), which was adapted from Wallace and Gutzler (1981).

pattern, (2) the Pacific/North American pattern, (3) the western Atlantic pattern, (4) the western Pacific pattern, and (5) the Eurasian pattern. The northern Pacific oscillation was identified early by Walker and Bliss (1932) and mentioned in Bjerkness (1969). A zonally symmetric, global-scale *see-saw* between polar and temperature latitudes, which is most clearly defined in the sea level pressure field, was pointed out by Lorenz (1951). An excellent discussion about the Southern Oscillation, El Niño and La Niña, was recently given by Philander (1990).

Figure 8.1 shows the western Atlantic pattern at 500 mb. Shown is the difference between the 10 months with the largest positive value of an index of pattern and the 10 months with largest negative values in a set of 45 winter months, as calculated by Wallace and Gutzler (1981).

It has been shown that the far-field response to large-scale forcing in the atmosphere is dominated by the external Rossby wave (Held 1983). Both the numerical model solutions and the analytical solutions by Hoskins and Karoly (1981), Karoly and Hoskins (1982), Karoly (1983), and Yang and Yang (1988, 1990) support this idea. It was found that the Rossby wave packet theory can provide a simple interpretation of the results from numerical model simulations and observations.

Yang and Yang (1988, 1990) studied the integral properties of wave packet by considering the background of the basic flow and topography, and obtained the condition for the stationary forcing wave packet in an asymmetric basic flow. They applied the idea to the southwesterly and the southeasterly jet. A concept of wave packet breaking and wave guide branching is proposed; according to the location of the wave packet with respect to the jet and the condition satisfied by the stationary wave packet, its trough (ridge) line has to be deformed, bending, breaking, and bifurcating. The wave guide hence bifurcates correspoondingly. As for the southwesterly jet, one bifurcated wave packet propagates toward high latitudes, and penetrates upwards into the stratosphere; the other wave packet propagates toward the low latitudes a far less distance than the first one, and only attains the tropopause. The Eurasian pattern teleconnection is investigated by means of the feature of the wave guide bifurcation of the Rossby wave packet in the southwesterly jet. This teleconnection is characterized by two positive correlation centers, one located down low latitudes, the other located down high latitudes, and 100 latitudes apart (Gambo and Kudo, 1983). In the asymmetric basic flow with topography similar to that in Section 3.4.2, the change in the wave packet energy can be stated as follows:

$$
\frac{\partial}{\partial T} \int\int_S \frac{1}{2} K^2 |\Psi_0|^2 = \int\int_S |\Psi_0|^2 mn \left(\frac{\partial U}{\partial Y} + \frac{\partial V}{\partial X} \right) dX dY
$$

$$
- \int\int_S \frac{|\Psi_0|^2}{2} \frac{\partial B}{\partial X} dX dY + \int\int_S \frac{|\Psi_0|^2}{2} K^2 \left(\frac{\partial U}{\partial X} + \frac{\partial V}{\partial Y} \right) dX dY
$$

$$
+ \int\int_S |\Psi_0|^2 \left(m^2 \frac{\partial U}{\partial X} + n^2 \frac{\partial V}{\partial Y} \right) dX dY \tag{8.1}
$$

from (3.82).

If the stationary Rossby wave packet is forced by topography or heating, we obtain the condition for such forcing stationary wave packets:

$$
\int\int_S mn|\Psi_0|^2 \left(\frac{\partial U}{\partial Y} + \frac{\partial V}{\partial X} \right) dX dY = \int\int_S \frac{|\Psi_0|^2}{2} \frac{\partial B}{\partial X} dX dY
$$

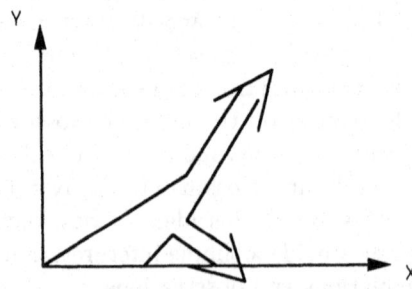

FIGURE 8.2. The breaking of the wave packet propagation passage in a southwesterly jet in the horizontal plane. After Yang and Yang (1990).

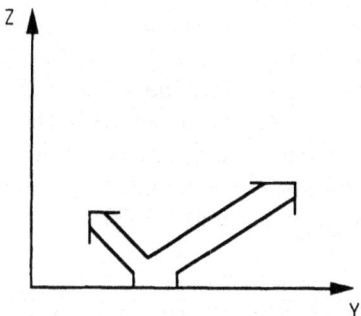

FIGURE 8.3. The breaking of the wave packet propagation passage in a south-westerly jet in the vertical plane, where Z_0 is the height of the tropopause. After Yang and Yang (1990).

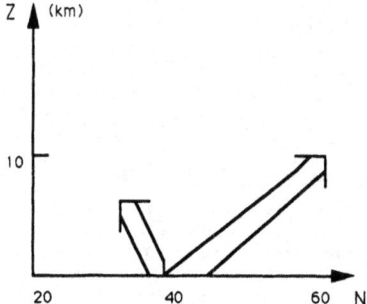

FIGURE 8.4. Ideal topographically forcing planetary wave propagation at $30°$ N. Adapted from Huang and Gambo (1983).

$$- \iint_S \frac{|\Psi_0|^2}{2} \left[(K^2 + 2m^2) \frac{\partial U}{\partial X} + (K^2 + 2n^2) \frac{\partial V}{\partial Y} \right] dX dY. \quad (8.2)$$

In a southwesterly jet, the condition becomes

$$\iint_S mn|\Psi_0|^2 \left(\frac{\partial U}{\partial Y} + \frac{\partial V}{\partial X} \right) dX dY < 0. \quad (8.3)$$

However, the change in the tilt of trough line of wave packet reads

$$\frac{D_g}{DT} \left(-\frac{n}{m} \right) = \frac{\partial U}{Y} - \frac{n^2}{m^2} \frac{\partial V}{\partial X} - \frac{1}{K^2} \frac{\partial B}{\partial Y}$$

$$+ \frac{n}{mK^2} \frac{\partial B}{\partial X} + \frac{n}{m} \left(\frac{\partial V}{\partial Y} - \frac{\partial U}{\partial X} \right). \quad (8.4)$$

The wave packet energy propagating down low latitudes must satisfy the following condition:

$$\frac{2B}{K^4} |mn| > V, \quad (8.5)$$

which means that only long wave energy is propagated into the low latitudes. In virtue of what was discussed in earlier chapters, we reach the conclusions of Yang and Yang (1988, 1990). Figure 8.2 and Figure 8.3 illustrate the bifurcation of the wave packet guides in the horizontal and vertical plane, where constant vertical energy propagation is assumed. Figure 8.4 shows the numerical results of topographically forcing wave propagation guide obtained by Huang and Gambo (1983).

8.3 Wave Packet Propagation in Zonal Flow (Hoskins and Karoly, 1981)

Consider the solutions of the linearized nondivergent barotropic vorticity equation on the sphere. The equation can be derived in the same manner as was done in Chapter 3. It is convenient to use a Mercator projection of the sphere (Phillips, 1973):

$$x = a\lambda \tag{8.6}$$

and

$$y = a \ln[(1 + \sin\phi)/\cos\phi], \tag{8.7}$$

where a is the average earth's radius. Then

$$\frac{1}{a\cos\phi}\frac{\partial}{\partial\lambda} = \frac{1}{\cos\phi}\frac{\partial}{\partial x}, \tag{8.8}$$

$$\frac{1}{a}\frac{\partial}{\partial\phi} = \frac{1}{\cos\phi}\frac{\partial}{\partial y}, \tag{8.9}$$

$$\nabla^2 = \frac{1}{\cos^2\phi}\left(\frac{\partial^2}{\partial x^2} + \frac{\partial^2}{\partial y^2}\right), \tag{8.10}$$

$$\cos\phi = \operatorname{sech} y/a, \tag{8.11}$$

and

$$\sin\phi = \tanh y/a, \tag{8.12}$$

where ϕ is the latitude of the earth, λ is the longitude, and a is the average radius of the earth. The Mercator basic zonal velocity

$$U = \bar{u}/\cos\phi \tag{8.13}$$

is proportional to the angular velocity. The equation for the horizontal stream function perturbation ψ, on multiplying by $\cos^2\phi$, takes the form

$$\left(\frac{\partial}{\partial t} + U\frac{\partial}{\partial x}\right)\left(\frac{\partial^2\psi}{\partial x^2} + \frac{\partial^2\psi}{\partial y^2}\right) + \beta_M\frac{\partial\psi}{\partial x} = 0, \tag{8.14}$$

where

$$\beta_M = \frac{2\Omega}{a}\cos^2\phi - \frac{d}{dy}\left[\frac{1}{\cos^2\phi}\frac{d}{dy}(\cos^2\phi U)\right] \tag{8.15}$$

is $\cos \phi$ times the meridional gradient of the absolute vorticity on the sphere. As before, we assume a solution with WKB form such as

$$\psi = A(X, Y, T)e^{\theta(X,Y,T)/\varepsilon}, \tag{8.16}$$

where

$$A(X, Y, T) = A_0(X, Y, T) + \varepsilon A_1(X, Y, T) + \varepsilon^2 A_2(X, Y, T) + \cdots, \tag{8.17}$$
$$(X, Y, T) = \varepsilon(x, y, t), \tag{8.18}$$

and

$$\varepsilon = \frac{\text{length scale of the perturbations}}{\text{length scale of the mean flows}} \ll 1. \tag{8.19}$$

Taking β_M as B_1, we can obtain the equations that govern the evolution of the wave packet, as we did in Section 2.5. Substituting the solution (8.16) into eq. (8.14), we obtain the zero order equation, that is, the dispersion relation,

$$\sigma = Um - \frac{\beta_M n}{m^2 + n^2}, \tag{8.20}$$

where σ is local frequency and m and n are local wave numbers in the X-direction and the Y-direction, respectively, as defined in earlier chapters.

The wave packet will be propagated along the group velocity as shown above. The group velocity is now

$$C_{gX} = \frac{\partial \sigma}{\partial m} = \frac{\sigma}{m} + \frac{2\beta_m m^2}{(m^2 + n^2)^2} \tag{8.21}$$

and

$$C_{gY} = \frac{\partial \sigma}{\partial n} = \frac{2\beta_M mn}{(m^2 + n^2)^2}. \tag{8.22}$$

However, the equations governing the structure of wave packet can be also derived from (8.20) with the method developed in Chapter 2, that is

$$\frac{D_g m}{DT} = 0, \tag{8.23}$$

$$\frac{D_g n}{DT} = -m\frac{\partial U}{\partial Y} + \frac{m}{(m^2 + n^2)^2}\frac{\partial \beta_M}{\partial Y}, \tag{8.24}$$

and

$$\frac{D_g \sigma}{DT} = 0. \tag{8.25}$$

Therefore, along the characteristic line, the direction of wave activity, m and σ, must be constant. However, the meridional wave number n will vary along the characteristic, as described by (8.24). Since $\partial U/\partial Y$ and

$\partial \beta_M / \partial Y$ depend on the latitude, that is, Y, the structure independence theorem stated in Chapter 2.4 no longer holds. Hence, we have to consider the propagation property of the wave packet in this problem. Moreover, teleconnections are found to be associated with the propagation property of the wave packet, especially of stationary wave packets. In the following, we study the propagation property of such wave packets.

To describe stationary wave phenomena, we consider the case $\sigma = 0$. Then the characteristic line, that is, the wave packet path, or ray, is given by

$$\frac{dY}{dX} = \frac{C_{gY}}{C_{gX}} = \frac{n}{m}. \tag{8.26}$$

Along the characteristic line,

$$m = \text{constant} \tag{8.27}$$

and

$$m^2 + n^2 = K_s^2, \tag{8.28}$$

where the stationary wave number is

$$K_s = \left(\frac{\beta_M}{U}\right)^{1/2}. \tag{8.29}$$

The magnitude of the group velocity from (8.21) and (8.22) is

$$C_g = 2\frac{m}{K_s}U. \tag{8.30}$$

Therefore, we conclude that on the Mercator projection or on the sphere, energy propagates along a characteristic line at a speed double that of the component of the basic flow in the direction of the characteristic line.

Let us first examine the case in which there is a constant angular velocity flow,

$$U = a\bar{\omega}. \tag{8.31}$$

From (8.15), we have

$$\beta_M = \frac{2\cos^2 \phi}{a}(\Omega + \bar{\omega}), \tag{8.32}$$

and the stationary wave number (8.29) is

$$K_s = (\varepsilon a)^{-1} \cos \phi, \tag{8.33}$$

where $\varepsilon^2 = \bar{\omega}[2(\Omega + \bar{\omega})]^{-1}$. From (8.26) and (8.28), the characteristic lines are given by

$$\frac{dY}{dX} = \left(\frac{K_s^2}{m^2} - 1\right)^{1/2}. \tag{8.34}$$

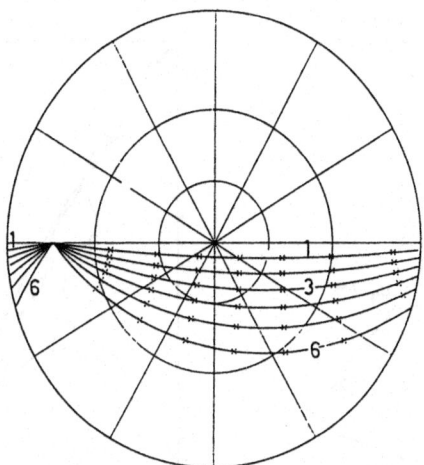

FIGURE 8.5. The wave packet paths and phases marked by a cross every 180° for a source at 15° N in a superrotation flow. If all wave lengths give a negative extremum at the source, the crosses mark the positions of successive positive and negative extreme. Lines of latitude and longitude are drawn every 30°, and the zonal wave numbers associated with paths are indicated. After Hoskins and Karoly (1981).

Substituting (8.28) into (8.34) and using the relation

$$\frac{d\phi}{d\lambda} = \cos\phi \frac{dY}{dX}, \tag{8.35}$$

we can integrate eq. (8.34) to obtain the characteristic equation

$$\tan\phi = \tan\alpha \sin(\lambda - \lambda_0), \tag{8.36}$$

where

$$\cos\alpha = \varepsilon am. \tag{8.37}$$

Equation (8.35) is the equation for a great circle through $\lambda = \lambda_0$ and $\phi = 0$ and reaching a latitude α, given by (8.37), which, from (8.33), is where $K_s = m$. Longuet-Higgins (1964a) was the first one to point out such great circles for characteristic lines. Equation (8.33) shows that the wave length on the sphere at the turning point of every characteristic line is ε times the planetary circumference. The speed of energy propagation in the great circle path on the sphere is $2\varepsilon am(a\bar{\omega})$, that is, a constant directly proportional to the zonal wave number. Thus, the time taken for energy to propagate 90° in longitude from the equation to the turning point is $(2\bar{\omega}am)^{-1}$ times the time for the zonal flow to travel 90°.

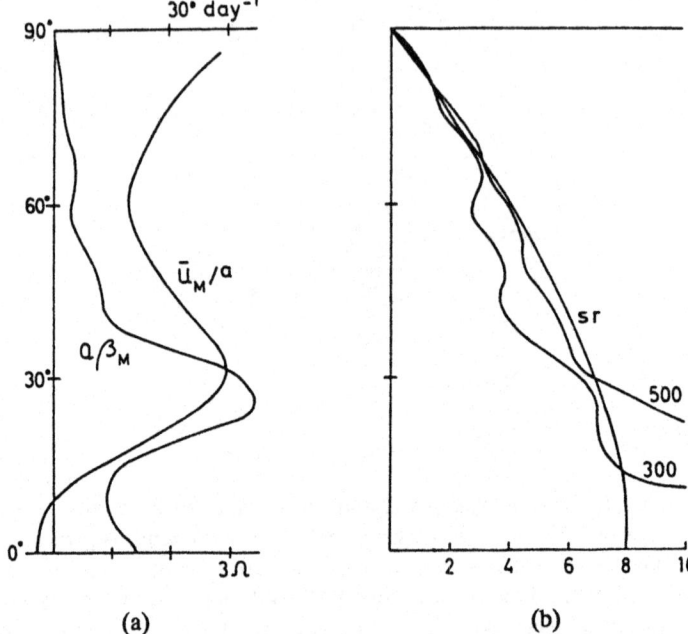

FIGURE 8.6. (a) U/a (degrees/day) and $a\beta_M$ for the Northern Hemisphere winter flow on 300 mb used in the model. (b) Stationary wave numbers aK_S for the superrotation flow, and the Northern Hemisphere winter flows on 300 mb and 500 mb. After Hoskins and Karoly (1981).

Applying the constant angular velocity (superrotation), used in Hoskins et al. (1977) and Grose and Hoskins (1979), to the model, that is, the speed at the equator was about 15 m/s, $\bar{\omega}/\Omega = 1/30.875$, and $\varepsilon \cong 0.125$, we obtain the following results. For wave number 1, the great circle reaches up to 83° but traveling at 3.8 m/s, the time taken for the wave activity to propagate from the equator to the turning point is about 31 days. For wave number 2, so that it is traveling twice as fast, it reaches 76° in about 15.5 days. The fastest possible energy propagation time around the sphere is 15.5 days. The wave length on the sphere at any turning point is about 5000 km. Our conclusions agree, to remarkable accuracy, with the previous findings by Hoskins et al. (1977) and Grose and Hoskins (1979), in which two different methods were used.

Suppose there is a source at 15°; according to the theory, we can calculate the results as shown in Figure 8.5. The wave packet paths (rays or ray paths) and phases are shown on a polar stereographic map. In this figure, characteristic lines and phases are marked by a cross every 180° for a source of 15° in a superrotation flow. However, for strict application of the theories, the zonal periodicity of the sphere should be dropped. Then the zonal wave number am would be a continuous variable and the picture

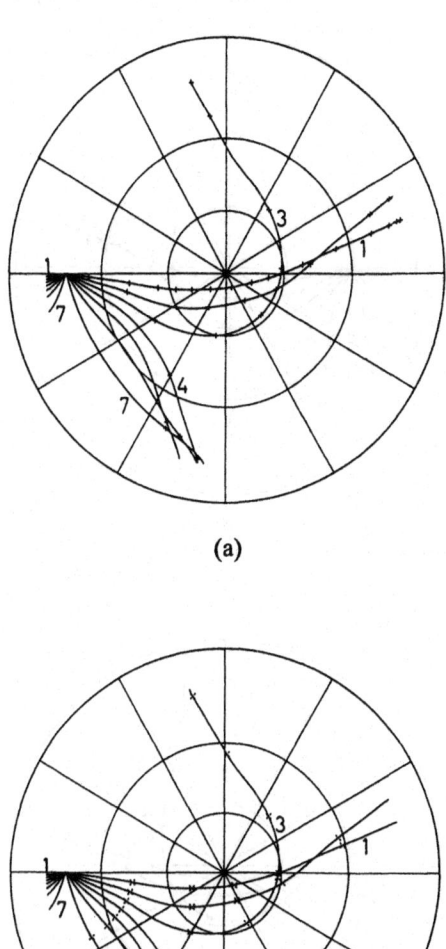

(a)

(b)

FIGURE 8.7. A source at 15° in the Northern Hemisphere winter zonal flow on 300 mb; (a) Paths and propagation are time marked by crosses every two days, and (b) wave packet paths and phases are marked every 180°. After Hoskins and Karoly (1981).

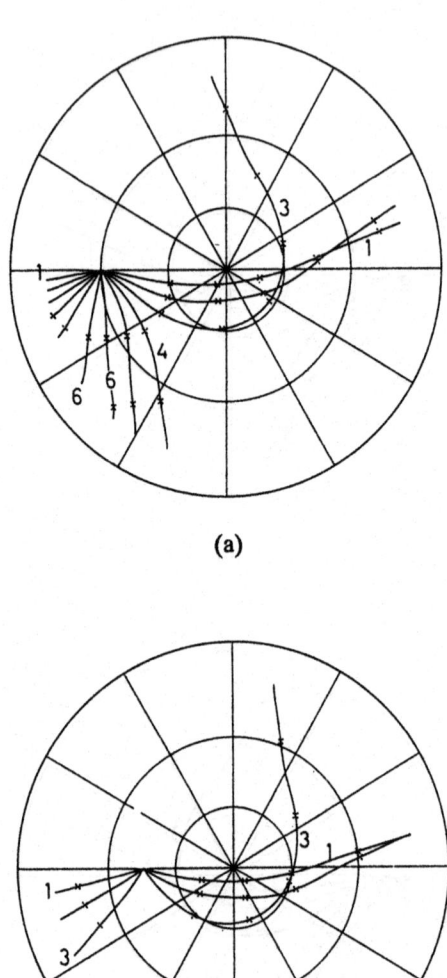

(a)

(b)

FIGURE 8.8. The wave packet paths and phases marked every 180° for (a) 30° and (b) 45° sources in the Northern Hemisphere zonal flow on 300 mb. After Hoskins and Karoly (1981).

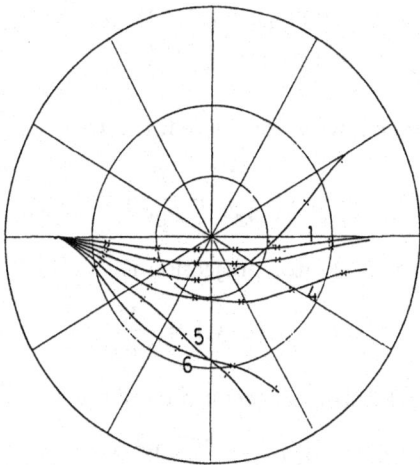

FIGURE 8.9. The wave packet paths and phases marked every 180° for a 15° N source in the Northern Hemisphere zonal flow on 500 mb. After Hoskins and Karoly (1981).

shown for integer values should be taken as a characteristic of the zonal wave numbers in a band centered on that value as a form of wave packet, as discussed in Chapter 2.

Hoskins and Karoly (1981) also presented the case for more realistic flows. Figure 8.6a gives the basic flows, $a\beta_M$ and aK_s, for the Northern Hemisphere winter on 300 mb. The stationary wave profile on 500 mb is also given in Figure 8.6b. The wave packet paths, energy propagation times, and phases for a 15° source are given in Figure 8.7, and the wave packet paths and phases for 30° and 45° sources are shown in Figure 8.8. As can be seen from Figure 8.7a, the wave packet with wave number 1 propagates from 15° into the 60° latitude circle within 9 days. For the wave packet with wave number 2, the time is only five days. It is rather interesting to note that zonal wave number 3 wave packet paths shown in Figure 8.8 for the basic 300-mb flow indicate the possibility of a wave packet propagating around a significant position of the hemisphere. Figure 8.9 shows the wave packet paths and phases every 180° for a 15° source in a Northern Hemisphere, 500-mb zonal flow.

The results demonstrate that linear barotropic models, baroclinic models, and wave packet theory analysis all give similar results, which are generally in agreement with those from observational studies and, to an extent, general circulation model integrations.

8.4 Wave Packet Propagation in Asymmetric Basic Flow (Karoly, 1983)

If the asymmetric basic flow is considered, the Laplacian is

$$\nabla^2 = \frac{1}{\cos^2 \phi} \left(\frac{\partial^2}{\partial x^2} + \frac{\partial^2}{\partial y^2} \right) = \frac{\nabla_M^2}{\cos^2 \phi}. \tag{8.38}$$

and the velocity in the Mercator projection is

$$\mathbf{V}_M = (U, V) = \frac{\mathbf{v}}{\cos \phi}. \tag{8.39}$$

The equation for the horizontal stream function ψ takes the form

$$\left(\frac{\partial}{\partial t} + U \frac{\partial}{\partial x} + V \frac{\partial}{\partial y} \right) \left(\frac{\nabla_M^2 \psi}{\cos^2 \phi} \right) + \frac{2\Omega \cos^2 \phi}{a} \left(\frac{1}{\cos^2 \phi} \frac{\partial \psi}{\partial x} \right) = Q, \tag{8.40}$$

where Q is the forcing of the mean flow. The equation for perturbation stream function can be written

$$\left(\frac{\partial}{\partial t} + U \frac{\partial}{\partial x} + V \frac{\partial}{\partial y} \right) \nabla_M^2 \psi' + B_1 \frac{\partial \psi'}{\partial x} + B_2 \frac{\partial \psi'}{\partial y} = 0, \tag{8.41}$$

where

$$B_1 = \bar{q}_y, \qquad B_2 = \bar{q}_x, \tag{8.42}$$

$$\bar{q} = \frac{\nabla_M^2 \bar{\psi}}{\cos^2 \phi} + 2\Omega \sin \phi, \tag{8.43}$$

is the absolute vorticity on the sphere, and

$$B_1 = \beta_M.$$

Again, by the use of the wave packet theory developed in Chapter 2, we can easily obtain the dispersion relationship as follows:

$$\sigma = Um + Vn + \frac{B_2 n - B_1 m}{m^2 + n^2}. \tag{8.44}$$

The group velocity is

$$C_{gX} = U + \frac{(m^2 - n^2)B_1 - 2mnB_2}{m^2 + n^2} \tag{8.45}$$

and

$$C_{gY} = V + \frac{2mnB_1 + (m^2 - n^2)B_2}{m^2 + n^2}. \tag{8.46}$$

The equations governing m and n are

$$\frac{D_g m}{DT} = -m \frac{\partial U}{\partial X} - n \frac{\partial V}{\partial X} + \frac{mB_{1X} - nB_{2X}}{(m^2 + n^2)^2} \tag{8.47}$$

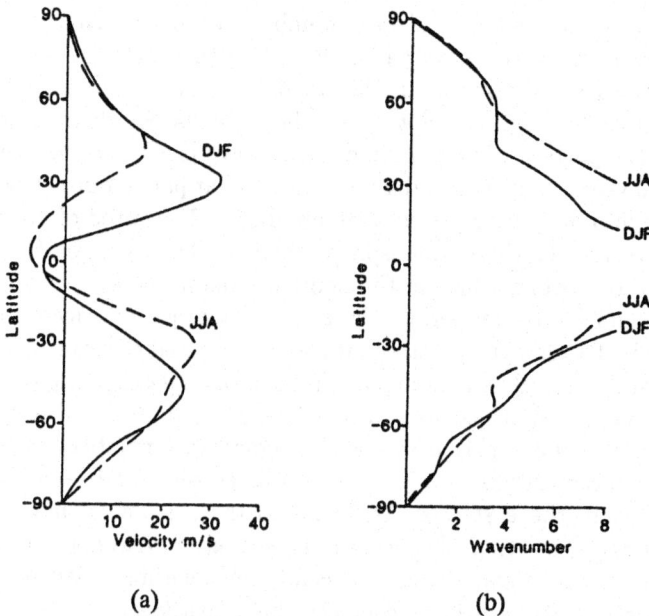

(a)   (b)

FIGURE 8.10. (a) Zonal flow for December to February (DJF) and June to August (JJA) used for the zonally symmetric basic state. (b) Stationary wave number profiles determined from the zonal flow. After Karoly (1983).

and

$$\frac{D_g n}{DT} = -m\frac{\partial U}{\partial Y} - n\frac{\partial V}{\partial Y} + \frac{mB_{1Y} - nB_{2Y}}{(m^2 + n^2)^2}, \qquad (8.48)$$

where again

$$\frac{D_g}{DT} = \frac{\partial}{\partial T} + \mathbf{C}_g \cdot \nabla   \qquad (8.49)$$

is the material derivative along the characteristic path.

Since B_{2Y}, B_{1X}, B_{2X}, and B_{1Y} still depend on the spatial variables, the structure independence theorem stated in Chapter 2.4 is not valid. However, what we are interested in now is the propagation properties and their implications in teleconnections. Therefore, for our assumed basic flow and waves, we can calculate the characteristic paths of the wave packet and propagation speed, as we did in the previous section.

The basic states are spherical harmonic representations of the global–zonal mean, 300-mb zonal flow for the seasons December to February (DJF) and June to August (JJA), from Newell et al. (1972). Figure 8.10 shows the flows and the associated stationary wave numbers profiles. The wave packet paths and propagation speeds for sources at 30°N and 30°S in DJF flow are shown in Figure 8.11 for both stationary and low-frequency waves. The low-frequency waves have periods of 50 and 20 days, which correspond to easterly phase speeds for zonal wave number 1 of 9.3 and 23.2 m/s, respectively. The paths of the wave packet for stationary waves in the Northern Hemisphere are the same as in the previous section, except that a rectangular latitude–longitude plot is used here. In the Southern Hemisphere summer, the pattern of propagation of the wave packet is much the same, with zonal wave numbers 2 and 4 following similar paths, but not propagating as far into high latitudes as wave number 1. The speed of propagation is faster in the Southern Hemisphere than in the Northern Hemisphere because of the stronger flow in the Southern Hemisphere.

The results for 50-day period waves and 20-day period waves are shown in Figure 8.11b and c, respectively. It should be noted that the paths of the wave packet for zonal wave number 1 and 2 have larger changes, with strong remarked westward propagation at low latitude and increased propagation speed. For the 20-day period waves, the higher wave numbers have greater meridional propagation into the equatorial easterly and large westward propagation in the easterlies. Wave numbers 1, 2, and 3 have westerly phase speeds higher than the westerly jet maximum and high propagation speeds, so they propagate freely between the Northern Hemisphere and the Southern Hemisphere, being reflected at high latitudes.

Results for the equatorial easterlies in the JJA flow could similarly be obtained (Karoly, 1983). The equatorial easterlies in the JJA flow are larger than in the DJF flow, so waves need larger easterly phase speeds before they can propagate between the two hemispheres.

The results so obtained agree with the findings of Hoskins et al. (1977). They found that, for realistic zonal flows with equatorial easterlies, large-scale transients and the stationary response are trapped poleward of the easterlies, in their studies of energy dispersion on the sphere, using a linearized barotropic number model.

Using a basic state with a zontally varying easterly jet at the equator observed in the tropical upper troposphere from Newell et al. (1972), as shown in Figure 8.12, we obtain the paths and speeds by integration. The zonal mean flow is a smoothed representation of the Northern Hemisphere, winter 300-mb flow, having a midlatitude westerly jet of 31 m/s and equatorial easterlies of 6 m/s. Results for stationary waves with zonal wave numbers 3 and higher are obtained. The paths of the wave packet for the two sources

---

FIGURE 8.11. The wave packet paths and propagation speed, shown by crosses at two-day time intervals, for sources at 30°N and 30°S in the DJK flow. Some of the zonal wave numbers associated with the paths are indicated. (a) Stationary waves and (b) 50-day and (c) 20-day period waves, with an easterly phase speed. After Karoly (1983).

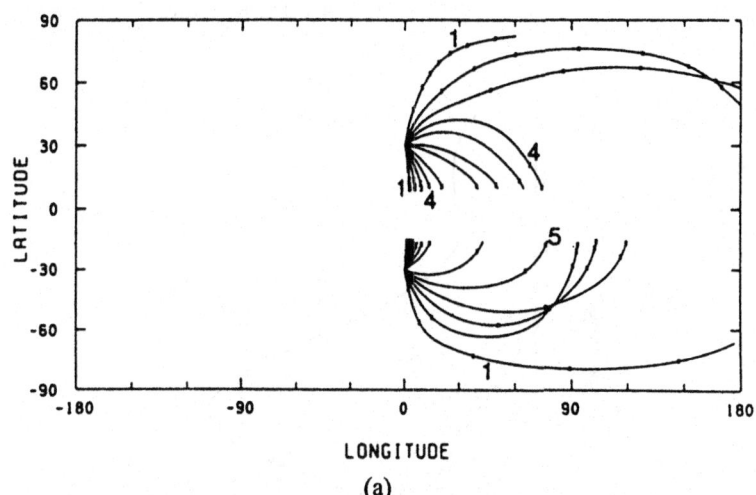

(a)

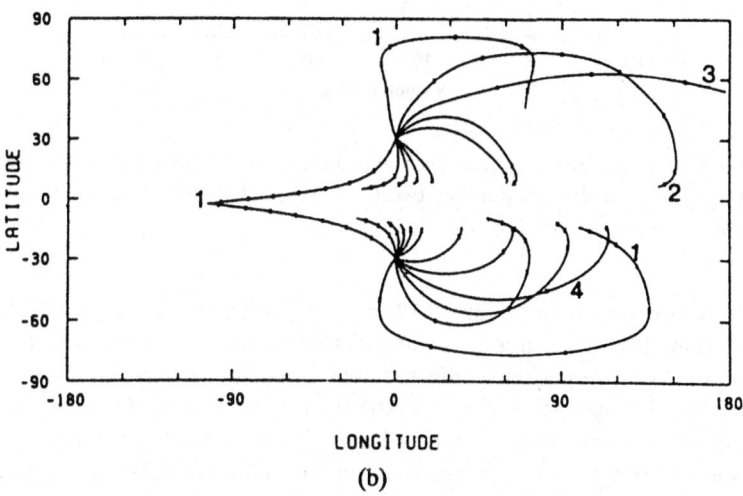

(b)

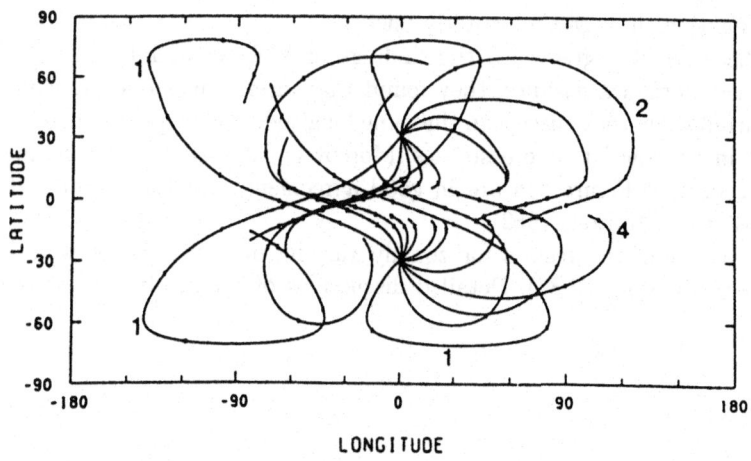

(c)

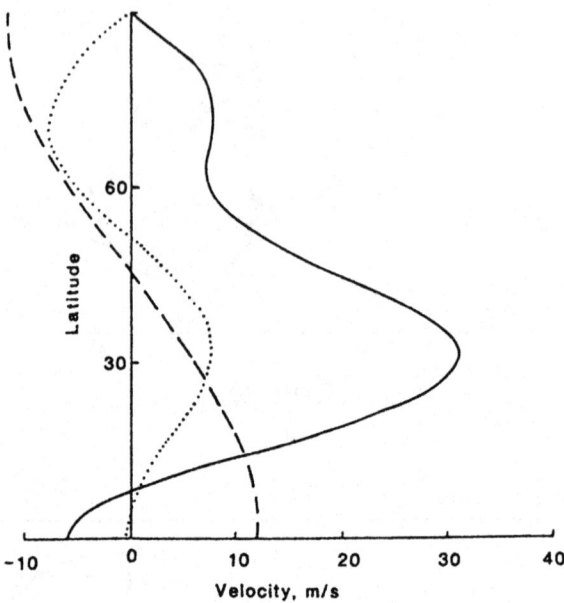

FIGURE 8.12. Zonal mean flow (—) and amplitude of the zonal wave number one (- - -) and zonal wave number two ($\cdots$) components for the zonally varying basic state. After Karoly (1983).

at $30°N$ are shown in Figure 8.13, together with the zonal flow. It was found that, for the source poleward of the tropical easterlies, the propagation of the wave packet is affected very little by zonal variation and the equatorward propagation is terminated at the critical line at low latitude. The speed of propagation shown in Figure 8.13a is small in the weak westerlies at low latitude. The phase variation along the paths, shown in Figure 8.13b, is similar to that for the zonally symmetric flow poleward of the cource but shows several amplitude maxima in the opposite hemisphere for propagation through the westerly duct.

These results agree with the findings of Webster and Holton (1982). In their numerical studies, they found that interhemispheric propagation occurred when the source is in the same longitude region as a westerly wind duct in the low latitude easterlies. Moreover, the positions of maxima in the results of Figure 11a are in good agreement with those obtained by Webster and Holton (1982).

In addition, the results for zonally varying mid-latitude jet were presented in Karoly (1983). Detailed discussions of the results can be found there.

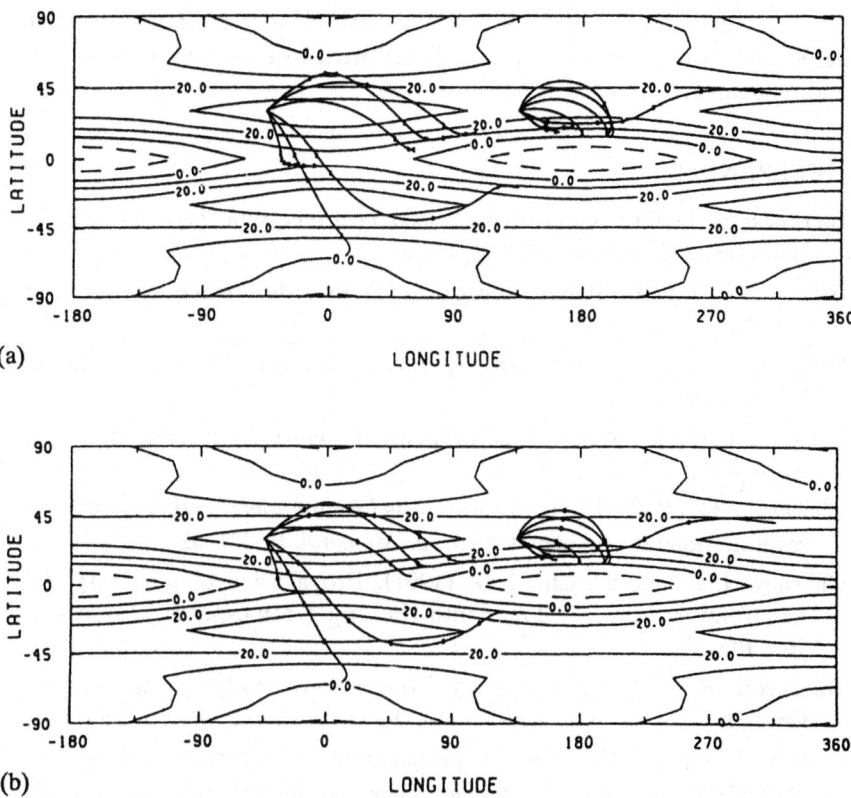

FIGURE 8.13. The wave packet paths for stationary waves in the basic state with zonally varying low-latitude easterlies from sources at $30°N$ for zonal wave numbers greater than 3. Contours show the zonal flow, with a 10 m/s contour interval; dashed contour for easterly flow. (a) The propagation time is marked by crosses at 2-day intervals. (b) The phase variation is marked by circles at $180°$ intervals. If all wave lengths give an extremum at the source, the circles indicate the positions of successive extrema of opposite sign. After Karoly (1983).

8.5   Discussion

The wave packet theory is an effective and useful method for studying the propagation of forcing planetary waves in different basic states and provide a good guide to predicting the general pattern of perturbations induced by large-scale forcing, especially for stationary wave packets. However, caution should always be exercised before the theory can be directly applied to any real physical problem; in particular, its sensitivity and its limitation for a particular physical problem should be considered. Even for the teleconnection problem, further investigations are needed before the results can be directly interpreted in terms of teleconnection. Karoly and Hoskins (1982) extended their results into the three-dimensional propagation problem of wave packets with a zonally symmetric basic flow. They found excellent agreement, using the wave packet theory with results from observational and modeling studies.

REFERENCES

Bjerknes, J. (1969). Atmospheric teleconnections from the equatorial Pacific. *Mon. Wea. Rev.* **97**, 162–172.

Brillouin, L. (1946). *Wave Propagation in Periodic Structures*. McGraw-Hill, London.

Brillouin, L. (1953). *Wave Propagation in Periodic Structures*, 2nd ed. Dover, New York.

Brillouin, L. (1960). *Wave Propagation and Group Velocity*. Academic Press, New York.

Charney, J.G. (1949). On a physical basis for numerical prediction of large-scale motions in the atmosphere. *J. Meteorol.* **6**, 371–385.

Charney, J.G., and Drazin, P.G. (1961). Propagation of planetary-scale disturbances from the lower into the upper atmosphere. *J. Geophys. Res.* **66**, 38–109.

Dickinson, R.E. (1968). Planetary Rossby waves propagating vertically through weak westerly wind waveguides. *J. Atmos. Sci.* **25**, 984–1002.

Dickinson, R.E. (1969). Vertical propagation of planetary Rossby waves through an atmosphere with Newtonian cooling. *J. Geophys. Res.* **74**, 929–938.

Gambo, K., and Kudo, K. (1983). Three-dimensional teleconnections in the zonally asymmetric height field during the Northern Hemisphere winter. *J. Meteorol. Soc. Japan* **61**, 36–50.

Grose, W.L., and Hoskins, B.J. (1979). On the influence of orography on large-scale atmospheric flow. *J. Atmos. Sci.* **36**, 223–234.

Hayashi, Y. (1981). Vertical-zonal propagation of a stationary planetary wave packet. *J. Atmos. Sci.* **40**, 1197–1205.

Held, I.M. (1983). Stationary and quasi-stationary eddies in the extratropic troposphere: Theory. In *Large-Scale Dynamical Processes in the Atmosphere*, B.J. Hoskins and R. Peace, Eds. Academic Press, New York.

Holton, J.R. (1972). Waves in the equatorial stratoposphere generated by tropospheric heat sources. *J. Atmos. Sci.* **29**, 368–375.

Hoskins, B.J., and Karoly, D.J. (1981). The steady linear response of a spherical atmosphere to thermal and orographic forcing. *J. Atmos. Sci.* **38**, 1179–1196.

Huang, R., and Gambo, K. (1983). The response of an atmospheric multi-level model to forcing by topography and stationary heat sources in summer. *J. Meteorol. Soc. Japan* **61**, 495–509.

Karoly, D.J. (1983). Rossby wave propagation in a barotropic atmosphere. *Dyn. Atmos. Oceans*, **7**, 111–125.

Karoly, D.J., and Hoskins, B.J. (1982). Three-dimensional propagation of planetary wave. *J. Meteorol. Soc. Japan* **60**, 109–123.

Longuet-Higgins, M.S. (1964a). Planetary waves on a rotating sphere I. *Proc. Roy. Soc. London* **279**, 446–473.

Longuet-Higgins, M.S. (1964b). On group velocity and energy flux in planetary wave motions. *Deep-Sea Res.* **11**, 35–42.

Longuet-Higgins, M.S. (1965). Planetary waves on a rotating sphere II. *Proc. Roy. Soc. London* **284**, 40–68.

Longuet-Higgins, M.S. (1968). Double Kelvin waves with continuous depth profiles. *J. Fluid Mech.* **34**, 49–80.

Lorenz, E.N. (1951). Seasonal and irregular variations of the Northern Hemisphere sea-level pressure profile. *J. Meteorol.* **8**, 52–59.

Matsno, T. (1970). Vertical propagation of stationary planetary waves in the winter Northern Hemisphere. *J. Atmos. Sci.* **27**, 871–883.

Newell, R.E., Kidson, J.W., Vincent, D.G., and Boer, B.I. (1972). *The General Circulation of the Tropical Atmosphere and Interactions with Extratropical Latitudes*, Vol. I. MIT Press, Cambridge, Mass.

Philander, S.G. (1990). *El Niño, La Niña, and the Southern Oscillation.* Academic Press, New York.

Phillips, N.A. (1973). Principles of large-scale numerical weather prediction. In *Dynamical Meteorology*, P. Morel, Ed. Reidel, Hingham, Mass., pp. 1–96.

Rayleigh, Lord (1877). *Theory of Sound*, Vol. I. Macmillan Company, London. Reprinted 1945, Dover, New York.

Rhines, P.B. (1969). Slow oscillation in an ocean of varying depth. Part I: Abrupt topography. *J. Fluid Mech.* **37**, 161–189.

Rossby, C.G. (1945). On the propagation of frequencies and energy in certain types of oceanic and atmospheric waves. *J. Meteorol.* **2**, 187–204.

Rossby, C.G. (1949). On the dispersion of planetary waves in a barotropic atmosphere. *Tellus* **1**, 54–58.

Schopf, P.S., Anderson, D.L.T., and Smith, R. (1981). Beta-dispersion of low-frequency Rossby waves. *Dyn. Atmos. Oceans* **5**, 187–214.

Skovgaard, O., Johsson, I.G., and Bertelson, J.A. (1975). Computation of wave heights due to refraction and friction. *Proc. Am. Soc. Civ. Engr., J. Waterways, Harbors and Boastal Engin. Div.* **101** (WWL), 15–32.

Skovgaard, O., Johsson, I.G., and Bertelson, J.A. (1976). Computation of wave heights due to refraction and friction (closure). *Proc. Am. Soc. Civ. Eng., J. Waterways, Harbors and Coastal Engin. Div.* **102** (WWL), 100–105.

Smith, R. (1971). The ray paths of topographic Rossby waves. *Deep-Sea Res.* **18**, 477–483.

Tolstoy, I. (1973). *Wave Propagation.* McGraw-Hill, New York.

Trefethen, L.N. (1982). Group velocity in finite difference schemes. *SIAM Rev.* **23**, 113–136.

Tung, K.K., and Lindzen, R.S. (1979a). A theory of stationary long waves. Part I: A simple theory of blocking. *Mon. Wea. Rev.* **107**, 714–734.

Tung, K.K., and Lindzen, R.S. (1979b). A theory of stationary long waves. Part II: Resonant Rossby waves in the presence of realistic vertical shears. *Mon. Wea. Rev.* **107**, 735–756.

Vichnevetsky, R. (1981). Energy, group velocity in semi-discretizations of hyperbolic equations. *Math. Comp. Simulation* **23**, 333–343.

Vichnevetsky, R. (1984a). The energy flow equation. *Math. Comp. Simulation* **26**, 93–101.

Vichnevetsky, R. (1984b). The mathematics of energy propagation in numerical approximations of hyperbolic equations. In *Advances in Computer Methods for Partial Differential Equations V*, R. Vichnevetsky and R.S. Stepleman, Eds. IMACS, New Brunswick, N.J., 133–166.

Vichnevetsky, R. (1987). Wave propagation and reflection in irregular grids for hyperbolic equations. *Appl. Numer. Math.* **3**, 133–166; or In *Numerical Fluid Dynamics*, R. Vichnevetsky, Ed. North-Holland, Amsterdam.

Walker, G.T., and Bliss, E.W. (1932). World Weather V. *Mem. Roy. Metero. Soc.* **4**, 53–84.

Wallace, J.M., and Gutzler, D.S. (1981). Teleconnections in the geopotential height field during the Northern Hemisphere winter. *Mon. Wea. Rev.* **109**, 784–813.

Webster, P.J., and Holton, J.R. (1982). Cross equatorial response to middle-latitude forcing in a zonally varying basic state. *J. Atmos. Sci.* **39**, 722–733.

Yang, H., and Yang, D. (1988). Jet stream and the stationary forcing Rossby wave packet in relation to the teleconnection in the atmosphere. *Acta Meteor. Sinica* **46**, 403–411.

Yang, H., and Yang, D. (1990). Forced Rossby wave propagation and teleconnections in the atmosphere. *Acta Meteor. Sinica* **4**, 18–26.

Yeh, T.C. (1949). On energy dispersion in the atmosphere. *J. Meteorol.* **6**, 1–16.

Author Index

Subject Index

Applied Mathematical Sciences
